THE BUSINESS CASE FOR GIS

MEASURING UP

Christopher Thomas
Brian Parr
Britney Hinthorne

Esri Press
REDLANDS|CALIFORNIA

Esri Press, 380 New York Street, Redlands, California 92373-8100
Copyright © 2012 Esri
All rights reserved. First edition 2012
16 15 14 13 12   1 2 3 4 5 6 7 8 9 10

Printed in the United States of America

Library of Congress Cataloging-in-Publication Data
Thomas, Christopher, 1963 Aug. 6–
Measuring up : the business case for GIS / Christopher Thomas and Milton Ospina.
   p. cm.
   Includes bibliographical references.
   ISBN 1-58948-088-0 (pbk. : alk. paper)
   1. Decision support systems. 2. Geographic information systems. 3. Industrial management. I. Ospina, Milton. II. Title.
HD30.213.T47 2004
658.4'038—dc22   2004013898

Ask for Esri Press titles at your local bookstore or order by calling 800-447-9778, or shop online at www.esri.com/esripress. Outside the United States, contact your local Esri distributor or shop online at www.eurospanbookstore.com/esri.

Esri Press titles are distributed to the trade by the following:

In North America:
Ingram Publisher Services
Toll-free telephone: 800-648-3104
Toll-free fax: 800-838-1149
E-mail: customerservice@ingrampublisherservices.com

In the United Kingdom, Europe, Middle East and Africa, Asia, and Australia
Eurospan Group
3 Henrietta Street
London WC2E 8LU
United Kingdom
Telephone: 44(0) 1767 604972
Fax: 44(0) 1767 601640
E-mail: eurospan@turpin-distribution.com

Contents

v   Foreword

vii   Acknowledgments

ix   Introduction

2   Chapter 1: Save time

4   GIS keeps information flowing to response agencies and public during Queensland flooding

8   GIS delivers results to the Alameda County Registrar of Voters

10   Louisiana Army National Guard deploys GIS to make the most of its data

14   Chapter 2: Save money

16   GIS erases the cost of graffiti in Riverside, California

20   GIS-assisted permitting saves Honolulu money and generates new revenue

22   Saving livestock saves millions

26   Chapter 3: Avoid costs

28   Baltimore County, Maryland, strategic plans reveal millions in return on investment from GIS

32   Bonner County, Idaho, manages invasive weeds with GIS

34   Los Angeles Bureau of Sanitation uses GIS to avoid costs and improve workflows

36   Chapter 4: Increase accuracy

38   Charting the roads that connect the vast Navajo Nation

44   City of Alpharetta, Georgia, uses GIS to get an accurate census count

46   Hudson, Ohio, increases government transparency and workflow accuracy using GIS

48   Chapter 5: Increase productivity

50   GIS-based asset management peaks productivity for Colorado Springs

54   Collaborative Utility Exchange maps Johnson County utilities

56   GIS-based work order management increases productivity for the Consolidated Utility District

58   **Chapter 6: Generate revenue**

60   Pueblo County, Colorado, grows economy with GIS

64   Westfield, Indiana, gets fair revenues from GIS

66   Woodstock, Georgia, adds revenue by using GIS to analyze storm water billing

68   **Chapter 7: Increase efficiency**

70   Golden, Colorado, does more with less by using GIS for sign management

74   Irish councils automate data collection with GIS to improve public feedback processes

78   Mobile GIS improves code enforcement services in McAllen, Texas

80   **Chapter 8: Automate workflows**

82   The Virginia Department of Forestry uses GIS to automate workflows in office and the field

86   EastLink Tollway: GIS puts road construction project in the fast lane

90   Nova Scotia Power saves $200,000 in labor costs through GIS field connectivity

92   **Chapter 9: Manage resources**

94   Masdar City relies on GIS to help create one of the world's most sustainable urban developments

100   Mackay Regional Council shares data resources and increases productivity with web-based GIS

102   Glynn County, Georgia, uses mobile GIS to manage resources and lower costs

104   **Chapter 10: Aid budgeting**

106   Adams County, Illinois, uses GIS to rapidly assess flood damages

110   The City of Moreno Valley, California, analyzes foreclosures using GIS

112   The City of Redlands, California, uses GIS to budget services and cover new costs

115   **Summary**

117   **Case study credits**

Foreword

Governments around the world continue to value the role geography plays in improving our understanding of the impacts our decisions and actions have on our environment and economic stability, and in building sustainable government services and more livable conditions. This understanding of geography is now being extended to improve our interactions between governments and citizens by bringing the issues of our times into a context that is more meaningful to individuals themselves.

As we continue to improve the quality of our decisions and services, one technology stands out as a common denominator for improving the way we identify issues and how we model, analyze, collaborate, and communicate what we learn. That technology is geographic information systems (GIS).

Today, GIS is used by nearly every government discipline, including urban planning, public works, emergency management, housing, health and human services, environmental and natural resource management, and executive oversight of public policy. Citizens are, for the first time, engaging more with their governments through GIS-based transparency and accountability applications ranging from mobile citizen reporting to inclusive redistricting to communicating how taxpayer dollars are being spent.

While the acceptance of GIS is widespread, it is important to celebrate the return on investment realized by implementing GIS. *Measuring Up: The Business Case for GIS*, Volume 2, presents case studies about how state, regional, and local governments across the globe are integrating GIS into their daily business processes. *Measuring Up* provides insight into the innovation that comes from looking at problems spatially and the value GIS brings to improving efficiencies, decision making, planning, communication, and collaboration while creating transparency. In essence, the book shows the role GIS plays in supporting a government that is more accountable to the world we impact and the citizens we serve.

Jack Dangermond
President, Esri

Acknowledgments

Many thanks to the agencies and people who contributed their stories to this book. From their input, we have compiled thirty unique case studies from thirty unique locations, all with one thing in common—a clear and strong return on investment from the use of GIS.

For their contributions, input, and time, the authoring team would like to thank the following: Salvador Aguilar, Jr., Los Angeles Bureau of Sanitation; Joye Dell Baker, Adams County, Illinois; Eric Becker, City of Westfield, Indiana; Jonah Begay, Navajo Division of Transportation; Jason Clark, Thiess Pty. Ltd.; Colette Cronin, Dún Laoghaire-Rathdown County Council, Ireland; Tim Dupuis, Alameda County, California; Hazel Farley, Fingal County Council, Ireland; Chris Gerecke, Timmons Group; Derek Gliddon, Masdar; Nick Hancox, New Zealand Animal Health Board; Nick Hutton, Data Transfer Solutions, LLC; Sandra Janson, Mackay Regional Council, Australia; Stephen Jarrett, City of Moreno Valley, California; P. Hunter Key, Glynn County, Georgia; Andy Koostra, Consolidated Utility District, Rutherford County, Tennessee; Paul Leedham, City of Hudson, Ohio; Mike Liotta, Louisiana National Guard; Christopher Markuson, Pueblo County, Colorado; Terry McNabb, AquaTechnex; Philip Mielke, City of Redlands, California; Adam Montgomery, City of Alpharetta, Georgia; Emily Norton, City of Woodstock, Georgia; Jose J. Peña, City of McAllen, Texas; Quintin Pertzsch, City of Golden, Colorado; Shannon Porter, Johnson County, Kansas; Steve Reneker, City of Riverside, California; Andy Richter, City of Colorado Springs, Colorado; Traci Robinson, Johnson County, Kansas; Ken Schmidt, City and County of Honolulu, Hawaii; Brian Shannon, Nova Scotia Power Inc.; Ben Somerville, Esri Australia Pty. Ltd.; Rob Stradling, Baltimore County, Maryland; Steven Wardrup, City of Alpharetta, Georgia.

Special thanks also to the Esri Press staff for its professionalism and assistance throughout this project.

Introduction

Measuring the benefits of implementing GIS

GIS brings people from disparate groups and different disciplines together and helps them understand problems through a common visual language. GIS has demonstrated real business value and, as a result, numerous companies, agencies, and government organizations have established GIS programs during the last thirty years. The case studies documented in this book present the spectrum of benefits—tangible and intangible—that GIS brings to many disciplines and industries.

Many of the featured organizations are realizing a return on investment by integrating GIS into their information systems and maximizing the benefits of GIS by incorporating it into their daily business work flow. Organizations that implement GIS throughout all business tasks become more efficient, have access to more information, save time and money, and increase productivity as they create and maintain successful customer service programs. Communication improves internally and externally when a company uses GIS. Maximizing investment in GIS will lead to increased accuracy and will support the decision-making process within an organization—large or small.

Measuring Up: The Business Case for GIS, Volume 2, covers ten benefits of implementing GIS. The organizations featured have measured how GIS is making them more productive and efficient, how costs were avoided, and how the use of GIS technology has improved their communication internally and with their customers. This book shows how GIS can support organizations and businesses and help them realize benefits that include the following:

- saving time
- saving money
- avoiding cost
- increasing accuracy
- increasing productivity
- generating revenue
- increasing efficiency
- automating workflows
- managing resources
- aiding budgets

GIS is changing the way people interact with spatial information

In the three decades since the development of GIS technology, thousands of state and local governments, not-for-profits, corporations, and numerous other organizations have benefitted from integrating GIS software into their daily operations. The pioneering efforts of the federal, state, and local agencies that waded into the uncharted waters of this new science helped to define a vision that today is shared by many other business sectors.

What is the business value of adopting GIS? Initially, some departments saw GIS as an indispensable tool for their discipline, while others implemented GIS to achieve the goals of a specific project. Throughout the development of the technology, organizations have tested the power of GIS to enable them to keep pace with current customer expectations.

Increasingly, as these case studies show, organizations are realizing that GIS is a valuable tool to help them move forward efficiently while responding to issues of accountability and performance measurement.

GIS professionals have long had a vision of a societal GIS where private citizens and nongovernmental organizations would investigate, collaborate, and participate with government through spatially oriented data and applications. This vision is rapidly being realized by combining the strength of GIS, web, cloud, and smart mobile devices. Thus, GIS continues to improve the bond between government and the citizens it serves. As new applications emerge, GIS also will enable organizations to meet the challenges of reducing costs, delivering services faster, providing better customer service, and increasing productivity.

GIS has a positive ripple effect

Governments the world over first realized the benefits of GIS through the conversion of paper maps and records to digital spatial data. At this time, the benefits were tremendous, yet simple, as the spatially organized data could be referenced, reused, and exchanged at a pace greatly exceeding past capabilities.

A second wave of benefits was realized as more organizations began to use GIS within focused projects. During this phase, the method for geographic problem-solving matured and became more GIS-centric. A question was posed, data was gathered, analysis conducted, and results presented. Within each step, GIS played a critical role, but unlike analog methods of the past, this methodology produced data that could be reused, easily shared, or reapplied to test alternatives scenarios or answer other questions.

Seeing the tremendous benefits of GIS and how the geographic advantage could improve everything from common decision-making processes to the financial bottom line, organizations began to centralize their spatial databases, integrating them into enterprise-wide GIS programs that made GIS an integral component of their daily business workflows. In other words, GIS became invaluable as it was applied to problems and processes to achieve operational efficiency. Governments realized that success rates are increased and sustained by developing a solid foundation of data and services to apply across an entire organization to solve an array of problems. With a solid footing, a GIS can build on itself, and benefits will spill over from department to department.

Many of the case studies covered in this book show how GIS technology has become ingrained into a group's daily activities and how hard it is to imagine conducting business without GIS. Crime analysts are concerned not only with how much faster they can solve crimes with GIS but also that the overall crime rate has gone down. GIS is a vital tool for emergency managers because of the number of lives it helps save and for its capacity to supply critical information instantly. It also enables managers to study what happened during an emergency and model what-if scenarios, which aids in emergency planning and preparedness. Mobile government workforces in community development, public works, and health have leveraged GIS for routing and logistics to meet the demands to reduce their carbon footprints while improving productivity. Business units use GIS to help them increase revenue and because it helps them make their business processes more efficient.

As you read these pages, you are encouraged to visualize how GIS is benefiting or could benefit your organization and those you serve. The case studies featured show how investing in GIS is a solid business strategy, while providing a common language for discussion that brings stakeholders together in the decision-making process.

THE BUSINESS CASE FOR GIS

MEASURING UP

Chapter 1

Save time

Time management is a constant factor for improving service levels or increasing customer satisfaction. Introducing GIS into workflows or allowing the technology to automate processes tends to generate significant time savings by either speeding up process tasks or eliminating steps altogether. Besides saving labor hours and increasing turnaround time on tasks and projects, GIS benefits an organization's ability to do jobs that would otherwise be shifted to overtime, delayed, or left undone. Organizations that apply georeengineering efforts to increase efficiency or productivity can accomplish more tasks in the same period.

GIS applications free staff time and shift workloads. Analyses with GIS range from performing multiple "what-if" scenarios to automating entire manual processes. With GIS, tasks that took hours, weeks, or months to complete are finalized in minutes. Today's GIS leverages web-centric applications and mobile smart devices to yield government on-demand and on-the-job workforce applications.

GIS has proven itself as a success in time management. It helps governments meet or exceed projected timelines or lessen the time required to complete traditional workloads, despite the perception that there is never enough time to move an organization forward or increase constituent satisfaction.

GIS keeps information flowing to response agencies and public during Queensland flooding

SECTOR: Emergency response
Ben Somerville, Esri Australia Pty. Ltd.

Three-quarters of the State of Queensland in northeastern Australia was declared a disaster zone as a result of flooding that began in late December 2010. Flooding endangered residents and inundated homes, businesses, and farmland. Agencies responding to the disaster and residents of affected areas needed quick access to current information on the evolving situation. To meet this demand, a small GIS team in the capital city of Brisbane worked on behalf of the Brisbane City Council (BCC) to develop a GIS solution that shaved forty-eight hours off the time it took to deploy an interactive and online situational awareness map for the rapidly escalating emergency response operations.

Queensland flooding creates rapid demand for GIS

Queensland residents are familiar with moderate seasonal flooding in many rural areas during the wet summer months, but the flooding that began in December 2010 quickly rose to levels unlike anything on modern record. The rapid progression in the extent and the intensity of flooding put emergency response agencies on high alert and created an immediate demand for situational awareness maps to communicate flood extents, shelter sites, and road closures. They also needed this information all within a common platform that could be shared among multiple response agencies.

There was no doubt that the agencies working the disaster needed an online GIS to provide critical information about the rapidly changing ground conditions. They had to develop and deploy a mission-critical GIS on a dedicated web server immediately, and to do so during a period of extreme regional duress. This challenge was clearly summarized by Ben Somerville of Esri Australia Pty. Ltd., who stated, "Floods don't wait for you to get your systems up and running."

GIS is rapidly deployed with cloud technology

The BCC found an answer to its geospatial needs in cloud GIS, a tightly integrated system of GIS technologies that can be deployed off-premises within a short amount of time and administered remotely. By using a cloud-based GIS solution, Somerville and colleague Nick Miller were able to rapidly construct a web-mapping application in response to the rising demand for spatial information by all of the response agencies. The GIS made it possible for the team to quickly deploy critical maps and data to those who desperately needed them to keep pace with the deteriorating ground conditions.

> The GIS made it possible for the team to quickly deploy critical maps and data to those who desperately needed them to keep pace with the deteriorating ground conditions.

Somerville and Miller worked through the night on behalf of the BCC to get a Flood Common Operating Picture (COP), with GIS at its core, up and running with the latest, most accurate information. "We were asked to create the application at 7:30 Tuesday night," Somerville said. "We went to the crisis room, gathered the data, and were back at the office an hour and a half later. Moving the data via FTP [file transfer protocol] to the cloud took the longest—nearly six hours. While that was happening, we created a local version of the COP app that we could move to the cloud. At 7:00 a.m. Wednesday, the web-mapping application was operational."

Team members working on the application saved a considerable amount of time by leveraging the cloud to build their GIS solution. No time was lost tracking down, setting up, and configuring a web server for the online GIS application because all of the IT resources they needed were already available to them within Esri's cloud solution.

The cloud-based GIS also eliminated the troubling question of where to place the hardware for a mission-critical application when most of the normal options were being cut off by flooding. The cloud servers were located nowhere near the event, yet they were still readily accessible by the GIS team working on the Flood COP. All that Somerville and Miller needed to access the application server was an Internet connection; no time was lost figuring out where to place the server or how to access it.

GIS speeds information delivery and improves flood response

As the flooding was reaching epic proportions, so were response efforts on the ground, which included local teams from the BCC, Queensland Fire and Rescue Service, Queensland Police Service, and the Australian Defense Force. The Flood COP provided by the GIS

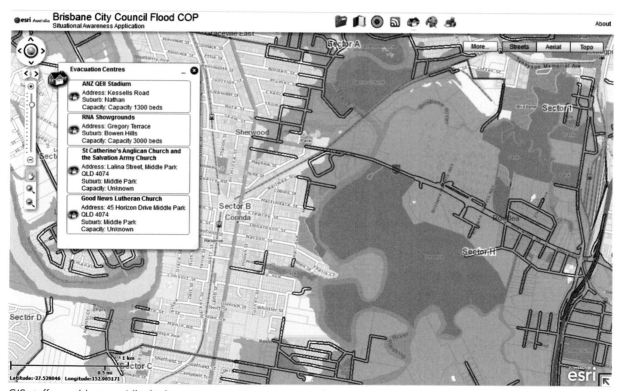

GIS staff was able to rapidly deploy a common operating picture for agencies responding to massive flooding in and around Brisbane by leveraging a cloud-based GIS solution. Courtesy of Esri Australia.

became invaluable to these agencies by giving them near-real-time feeds of rescue activities, current and predicted flood extents, and the spatial data and maps critical for coordinating relief and rescue efforts. It was no ordinary flood event and there was no time to wait for information. The real-time situational awareness provided by the GIS equated to better coordination between agencies and faster and more targeted response efforts on the ground.

Web traffic to the online GIS application rose dramatically after its URL link was made public. Within a few days after the link was posted, the site received over 3.5 million hits.

Web traffic to the online GIS application rose dramatically after its URL link was made public. Within a few days after the link was posted, the site received over 3.5 million hits. The dramatic jump in traffic made it apparent that the public was also demanding timely and accurate maps about the event. As Somerville recalled, "As soon as the application was published on the BCC website, it quickly went viral."

The BCC was able to leverage massive enterprise resources in the cloud to instantaneously handle the spike in demand as the hits to the GIS site soared. Had Somerville and Miller used a traditional in-house deployment instead of the cloud-based GIS, they would not have had the resources to pull together all the additional equipment and software in time to meet the overwhelming, and unexpected, demand placed on the server.

Somerville approximated that by going with a cloud-based GIS, they were able to save forty-eight hours in the time it took to get the Flood COP up and running—an estimated twenty-four hours saved in the retrieval and set up of hardware, plus another twenty-four hours saved

while installing, configuring, and testing the operating system and application software. With the intensity of the flooding quickly elevating, the time savings played a critical role in meeting the informational demands of all the overall response effort.

By going with a cloud-based GIS, they were able to save forty-eight hours in the time it took to get the Flood COP up and running.

The GIS-based Flood COP was a valuable public asset. It helped locals evacuate from harm's way while also helping people with friends and family in Queensland get information about the event. For example, Somerville knew of one business owner who used the timely information provided by the GIS to stay ahead of the flood and the associated road closures to safely remove over AU$100,000 of computer equipment from his building. This story was repeated many times over as residents and business owners used the GIS to determine what areas were going to flood so that they could take the appropriate action to get to safety.

Residents and business owners used the GIS to determine what areas were going to flood so that they could take the appropriate action to get to safety.

The combination of GIS with cloud technology was so successful to the region during the disaster that the BCC today maintains an ArcGIS for Server template of its Flood COP within the cloud. It is a perfect solution for incident-driven events like flood response because it essentially allows agencies to rapidly spin up the GIS applications they need, when they are needed, without having to make and maintain a large investment in hardware and software that, other than during the duration of the incident, will mostly lie dormant. As much as the BCC and its residents would rather not have to fire up the Flood COP again, it is good to know it is there in the cloud ready to provide critical GIS functionality on a moment's notice.

GIS and related technologies played a critical role during the massive flooding that struck Brisbane, Australia, in December 2010.
Courtesy of Esri Australia.

GIS delivers results to the Alameda County Registrar of Voters

SECTOR: Elections
Tim Dupuis, Alameda County Chief Technology Officer, Alameda County, California

While preparing for the next election in Alameda County, California, the county's director of information technology, Dave Macdonald, saw a clear need to modernize election workflows. Past elections were carried out effectively, but many of the processes were manual and too time-consuming. Responding to this, Macdonald initiated a project to improve election work by leveraging GIS and other technologies.

The Registrar of Voters (ROV) teamed with Weston Solutions, Inc. to develop a solution that built on the San Francisco Bay Area county's existing enterprise GIS. They incorporated the county's base layers directly into editing tasks, analysis operations, and online election maps. Specialized tools were then created for election staff to use for consolidating precinct boundaries based on specific population criteria, and GIS-based workflows were established for analyzing the suitability of building sites for new polling stations.

> Making the switch to GIS means precinct consolidation that traditionally took six election technicians up to three weeks to complete now takes three technicians one to three days to complete. Tim Dupuis, Chief Technology Officer, Alameda County

The new tools produced substantial savings by reducing the time needed to carry out critical election processes, especially in the areas of precinct consolidation and locating new polling stations. Speaking about this impact, the county's chief technology officer, Tim Dupuis, said, "Making the switch to GIS means precinct consolidation that traditionally took six election technicians up to three weeks to complete now takes three technicians one to three days to complete."

With any election comes a healthy demand for results. Knowing this, the ROV also built a public-facing website that automatically updates elections result maps in nearly real time through a direct link to the GIS. The application saves time for the public, media, and political organizations looking to quickly get results, and it also saves time for ROV personnel by reducing the number of requests they get from these same groups for election outcomes.

When asked if changing from a paper-based to a technology-based approach had a positive impact on the election day work, Dupuis stated, "We recorded some of our fastest times this year," making it clear that the switch was the correct choice for the ROV and the voters of Alameda County.

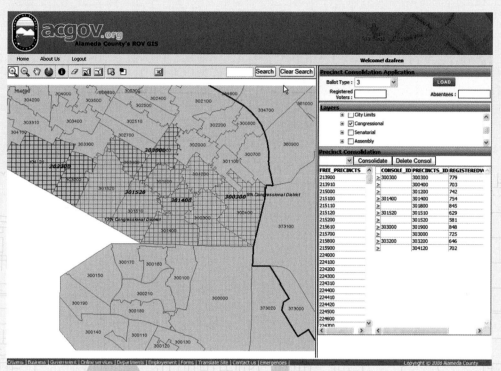

Tools, such as this one used to consolidate precinct boundaries into evenly populated zones, were built to assist election workflows. They were distributed via the county's intranet. Courtesy of Alameda County, California.

Louisiana Army National Guard deploys GIS to make the most of its data

SECTOR: Public safety
Mike Liotta, GIS Manager, Louisiana Army National Guard

From days to minutes, that was the time saved by the Louisiana Army National Guard (LANG) after it implemented a new GIS platform to manage its spatial data.

LANG stands ready to protect the safety and security of Louisiana's citizens when disaster arrives. It is called in to help when conditions are at their worst. Geospatial information is critical during these times, giving the guardsmen and guardswomen important information about what they are heading into as they set out to save lives and protect order.

After Hurricanes Katrina and Gustav, LANG realized future responses would benefit from more efficient methods for handling spatial data, especially imagery. Attaining the raw data was not the problem; in fact, it was plentiful. The challenge was making the volumes of data coming in from various sources rapidly available to LANG's operations.

During an event, large quantities of imagery and other spatial data come into LANG, such as this aerial photo of flooding. What was needed, though, was a solution for quickly operationalizing incoming data, making it available for situational awareness maps and field use. Courtesy of Louisiana Army National Guard.

Working to improve the process, LANG's GIS Manager, Mike Liotta, conducted a needs assessment that incorporated the lessons learned during previous responses along with input from the Joint Operations Center, which coordinates LANG's emergency operations and aviation units.

Liotta's assessment led to a new platform at LANG for distributing spatial data using ArcGIS for Server with the Image extension. This gave LANG the capability to quickly bring in, process, and then deploy large quantities of data. LANG also leveraged the development environment of ArcGIS for Server and built custom tools for interacting with the spatial data and maps.

LANG's web application permits users to access current imagery and data, make maps, and obtain information critical to its response efforts. Courtesy of Louisiana Army National Guard.

Soon after launching the new server, raster datasets that had been deemed unusable or only available through standard GIS software because of their size became easily accessible through simple web interfaces. These interfaces, in turn, provided critical response functions such as creating x,y coordinates from street addresses and constructing situational awareness maps for command staff.

> # When the next large-scale emergency occurs, we'll be able to quickly turn around newly acquired data and imagery to our command staff, soldiers, and first responders. This can make a big difference.
>
> Mike Liotta, LANG's GIS Manager

What once involved many hours of manpower and computer processing had evolved into a sleek, effective process for delivering imagery that fundamentally changed how LANG consumes and deploys large volumes of imagery. Comparing the old to the new system, Liotta explained, "With our previous methods, it would have taken several days to get the imagery in a widely accessible format." Now with ArcGIS for Server in place, it takes just thirty minutes to turn the raw data into a readily available image service. Liotta went on to say, "More importantly, when the next large-scale emergency occurs, we'll be able to quickly turn around newly acquired data and imagery to our command staff, soldiers, and first responders. This can make a big difference."

Chapter 2

Save money

The survival and success of an organization depend on many variables as management strives to become as efficient and productive as possible, maintain revenue streams, and control expenses. When income slows in a soft economy, management must quickly find ways to save money. Executives often turn to technology to help streamline or automate operations, and GIS has emerged as a technology with a reputation for cutting costs related to workflows or business problems.

GIS helps control spending through direct cost savings. GIS applications that improve decision making or increase productivity result in direct cost savings. For example, spatial analysis can guide decisions to select the optimal location for a fire station based on response times and property availability. It's a site-selection analysis, but one that typically saves money by leveraging GIS to find the least expensive property that can serve the most clients. The introduction of geo-enabled smart device applications now provides bidirectional data collection, reporting, and work order management. These field-to-office GIS solutions strip away redundant tasks, save time, and ultimately lead to significant cost savings. By using GIS to assist vehicle routing and logistics, organizations can reduce travel costs and in many cases reduce the number of fleet vehicles required to carry out operations, creating another level of savings.

GIS erases the cost of graffiti in Riverside, California

SECTOR: Public works and law enforcement
Steve Reneker, CIO, City of Riverside, California

Graffiti is a persistent challenge to cities and communities throughout Southern California, including Riverside, a city with a 300,000-plus population located sixty miles east of Los Angeles. In response to the challenge, the City of Riverside has developed a comprehensive GIS application that not only helps identify and prosecute the vandals who produce the graffiti, but also tracks the entire cost of the graffiti to the city.

Riverside budgets one million dollars annually for graffiti abatement programs. Each year, field crews working for the city respond to thousands of incidents. The city has learned that the prompt removal of graffiti (or tags) is a strong deterrent to additional tagging, and because of this, supports a graffiti abatement program that strives to remove all graffiti within twenty-four hours of it being reported.

To support their aggressive graffiti removal program, Riverside's Public Works, Police, and Information Technology Departments worked with Affiliated Computer Services (ACS) to develop a GIS application known as the Graffiti Abatement Tool (GAT) that coordinates interdepartmental efforts and addresses the problem of connecting multiple instances of graffiti to an individual vandal (or tagger). Today, Riverside's GIS application is reversing the prevalence of graffiti in the city and its associated costs by nearly $350,000 annually.

GAT was developed with ArcGIS for Server and relies heavily on a centralized geodatabase that is shared across the county. It stores and manages images of graffiti with other tabular data useful in tracking, prosecuting, and suing taggers for the city's costs to mitigate and remove the tags. By leveraging GIS technology, the application fosters the flow of information related to graffiti abatement from one city department to the next, while at the same time building a chain of evidence to catch vandals and prosecute them for the costs.

> Riverside's GIS application is reversing the prevalence of graffiti in the city and its associated costs by nearly $350,000 annually.

A snapshot from the Graffiti Abatement Tool (GAT) used by the City of Riverside. The tool uses GIS to improve how the city manages and shares information about graffiti incidents. Courtesy of City of Riverside, California.

Connecting workflows

The process begins when the public reports graffiti using Riverside's 3-1-1 system, either by calling in or sending an e-mail from a smart phone. Using the latter method, individuals photograph the graffiti with the camera on their phone, which automatically appends the latitude and longitude coordinates of the graffiti's location to the image, then they e-mail the information to 3-1-1.

Graffiti incidents reported to 3-1-1 are processed into work orders and sent to the Public Works Department, which then dispatches graffiti removal crews to the sites. Crews take pictures of the tag using a Ricoh GPS (Global Positioning System) camera and complete a customized digital form on the camera. This form records basic information about the incident and its removal. The images and data are then loaded onto a server that automatically adds the data to a spatial layer in the geodatabase.

After the Public Works Department enters new incidents into the GIS, an e-mail containing a new case number is generated and sent to the Police Department for investigation. Taggers sign their work with distinctive monikers, which makes it fairly simple to associate tags to a single vandal. Using GIS interfaces, the Police Department's graffiti task force examines the images and information about each incident and ties together those instances that are clearly from the same tagger. Additional information about the investigation and the cost to perform it are also entered into the system.

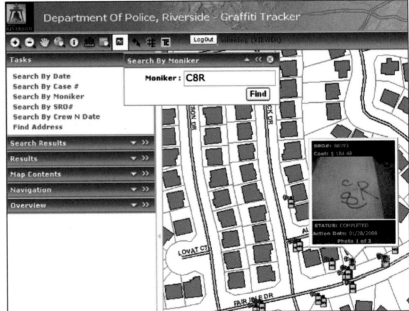

The work of a single tagger is identified at two separate locations. Investigators use GIS interfaces to associate multiple incidents of graffiti from different locations to a single vandal. Courtesy of City of Riverside, California.

The information that public works crews and police enter into GAT builds a strong base of evidence that is used by the city attorney in court cases against the vandals that are caught. The system provides the prosecutors with time-stamped pictures of the graffiti, the costs to remove and investigate the incidents, and the locations where the graffiti occurred—it's all in the GIS, ready to provide a complete picture of a tagger's crimes after he or she is apprehended.

Saving money and recouping costs

With the operational intelligence that GIS delivers to GAT, Riverside's Public Works Department can accurately track where and how much money is being spent to remove graffiti and determine where it is occurring the most. For example, when the public works abatement crew removes the graffiti, the cleanup method and materials used are entered, as well as the time required for removal. Using this information, the system then calculates the cost to remove the graffiti and stores it with the incident record. This helps the department better manage and budget its resources. From the overall improvements in graffiti management and the reduced time it takes to carry out standard workflows, such as locating graffiti in the field, Riverside's use of GIS to fight graffiti is saving the Public Works Department an estimated $200,000 a year in its graffiti abatement budget.

> Riverside's use of GIS to fight graffiti is saving the Public Works Department an estimated $200,000 a year in its graffiti abatement budget.

The Police Department contributes to GAT by building cases for prosecutors to go after vandals and recoup the costs caused by their damages. But the information and spatial data GIS returns to police are also very useful in other investigations. For example, taggers will often provide information about gang affiliation within their graffiti lettering, which can help police determine where gangs are active and to get a general idea of gang territory—something made possible by storing the graffiti data in a GIS.

The full effect of GAT plays out in court when the complete set of information is used against defendants. At each point along a graffiti incident's movement through GAT, the cost for its removal, investigation, and prosecution is recorded and saved with the record. Prosecuting attorneys use the system to quickly and objectively summarize all of these costs as they prepare their court case—costs that defendants are required to pay back to the city if they are found guilty. If the tagger is a minor, the city attorney can sue the parents or legal guardians to recoup the costs, giving them a strong incentive to keep their children away from such activities. By using this GIS-based system, Riverside is now recouping approximately $150,000 per year in graffiti costs through the successful prosecution of taggers.

> By using this GIS-based system, Riverside is now recouping approximately $150,000 per year in graffiti costs through the successful prosecution of taggers.

Widening the net

By using GIS to centralize and share information about graffiti that's tied to its real-world location, Riverside is recouping its costs to deal with graffiti and, more importantly, reducing the prevalence of this type of vandalism within its borders. The work to improve the system, however, goes on.

The organic nature of graffiti does not conform neatly, if at all, to jurisdictional boundaries. Considering this, Riverside hopes to widen GAT's net by sharing the system architecture with neighboring cities to build it into a regional graffiti/tagger tracking and abatement system. Other jurisdictions would simply provide their servers and GPS-enabled cameras to use the system, and ultimately the entire region could benefit through a reduction in the urban blight caused by taggers and their marks.

GIS-assisted permitting saves Honolulu money and generates new revenue

SECTOR: Planning
Ken Schmidt, GIS Administrator, City and County of Honolulu, Hawaii

Honolulu began using GIS in 1988, which put the city into a unique class of early adopters of the technology. Today, it still finds ways to improve government functions with GIS. A current example of this is a permitting application that saves the city thousands of hours in staff time each year, while at the same time bringing in hundreds of thousands of dollars in new revenue.

To complete a new permit application, applicants must answer several questions about the parcel where a proposed activity will occur. For example, city staff handling the application need to know the zoning, whether or not other permits have been issued at the site, if the parcel is affected by any regulatory issues, and who owns the property.

Honolulu's GIS holds the data layers required to answer these questions, and the staff members who processed the applications realized they could increase transaction rates by developing an online application that automatically checked a permit request against the spatial data stored in their GIS. Accelerating the process this way fit well with the mind-set of Honolulu's GIS manager, Ken Schmidt, who said, "We believe the GIS increases revenues through increasing the productivity of municipal operations."

> During the first two years, Honolulu averaged approximately $400,000 of new revenues from the system. HONLine also saved 32,000 hours in city staff time and roughly 17,000 hours in processing time for the permit applicants.

Recognizing the potential return on investment, administrative and permitting operation managers worked with the technical services groups to direct the setup of online services for several types of permits. The program is called HONLine and the results have been impressive. During the first two years, Honolulu averaged approximately $400,000 of new revenues from the system. HONLine also saved 32,000 hours in city staff time and roughly 17,000 hours in processing time for the permit applicants. Today, 25 percent of the city's building permits are issued online.

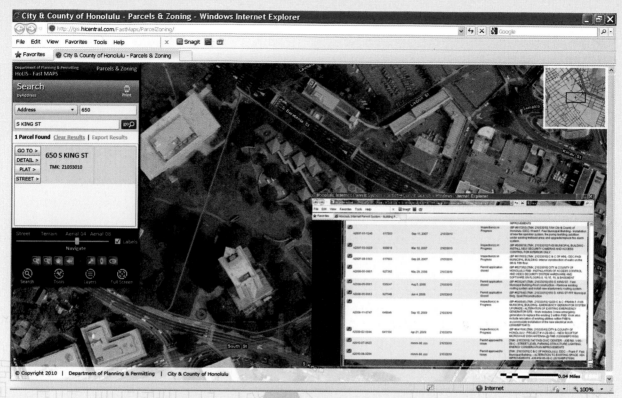

HONLine automates much of the question–answer process associated with permitting. For example, the application allows users to rapidly determine if any other permits have been issued at the parcel or building in question. Courtesy of City and County of Honolulu, Hawaii.

Considering the total amount of staff time saved and the new revenues generated, HONLine has a net positive impact of nearly one million dollars annually—an impact affirming Ken Schmidt's belief that, when wisely implemented, GIS produces positive returns.

First Year Return on Investment from HONline		
Internal	Revenue from permit submitted online	$425,000
	Saving in staff-hours (32,000 est. at $16.00 per hour)	$512,000
	Total:	$937,000
External	Total estimated time saved for permit applicants.	17,000 hours

HONLine, Honolulu's GIS-based permitting application, generates thousands annually in revenue and saves hundreds of hours in staff time. This table summarizes these returns on Honolulu's investment in HonLine. Courtesy of City and County of Honolulu, Hawaii.

Saving livestock saves millions

SECTOR: Environmental management
Nick Hancox, Operational Policy Manager, New Zealand Animal Health Board

Bovine tuberculosis (TB) is an infectious disease particularly damaging to the livestock and dairy industries. In New Zealand it poses a significant threat due to its primary vector, an invasive species of possum—the common brushtail—that spreads the contagion across the countryside and into cattle, dairy, and deer herds. It's a problem that New Zealand, well known for its meat and dairy exports, spends nearly NZ$80 million annually to mitigate and prevent.

In response to the severe risks and high costs associated with bovine TB and the ongoing need to control brushtail possum populations, New Zealand's livestock industries and government formed the Animal Health Board (AHB) and in 1998 entrusted it with the mission to protect the nation's beef, dairy, and deer livestock from the disease. Protecting the nation's livestock is a continual challenge, and today, AHB is achieving considerable success in its mission by using GIS to assist in its disease management practices.

> Protecting the nation's livestock is a continual challenge, and today, AHB is achieving considerable success in its mission by using GIS to assist in its disease management practices.

Prior to AHB, much of bovine TB activity (especially possum control) was managed independently by each of the country's regional councils. This decentralized approach led to differing management practices across the nation, with no easy way to measure or control activities or outcomes. The situation hindered AHB's ability to cost-effectively manage the disease at the national level, where it needed to achieve and demonstrate success.

In order to resolve these problems, AHB moved to bring all possum control planning and management in-house and also developed a centralized GIS to standardize the flow of information associated with its work. AHB partnered with Eagle Technology Group, Ltd., to build the GIS solution, which is known as VectorNet. This solution incorporated the best vector management practices from the regional councils, along with new planning, management, and data analysis functions to create a consistent, accurate, and easy-to-manage geodatabase used within all aspects of AHB's work to eradicate bovine TB from possums and other wildlife.

In the first decade alone, AHB estimates that the GIS-based application will save NZ$30 million in operating costs and completely pay for itself in fewer than four years.

Today, approximately forty AHB staff members use VectorNet for contract management, strategic planning, and reporting. Contractors update the database directly from the field with GPS-enabled handheld devices. Using statistical models to measure possum density and infection risk, the GIS generates exact locations within specific regions to implement control projects such as baiting or trap lines. This geospatial approach, integrated with operational data, creates verifiable processes to better manage current projects and formulate future predictions about where the disease will reemerge.

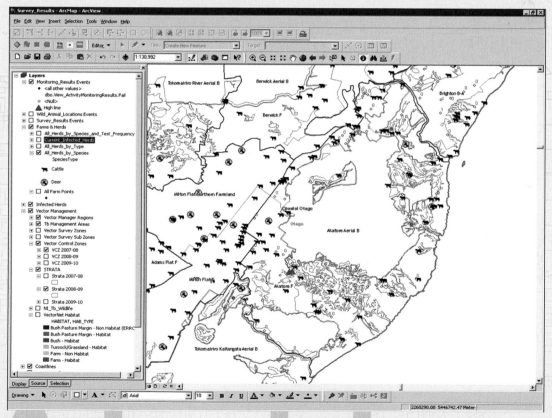

This is one of the GIS maps AHB uses to analyze livestock locations and disease rates within its management zones. GIS is giving AHB a significant advantage in the work to eliminate bovine TB. Courtesy of Nick Hancox, Operational Policy Manager, AHB.

VectorNet was, and continues to be, a great success. It's giving New Zealand and AHB the upper hand in their fight against bovine TB, while at the same time saving millions of dollars through better and more efficient management. In the first decade alone, AHB estimates that the GIS-based application will save NZ$30 million in operating costs and completely pay for itself within four years.

Discussing VectorNet's return on investment, AHB Chief Executive William McCook said, "That's just the beginning. For instance, we expect a 1 percent efficiency gain on the overall vector program budget through consistent, accurate, complete, and timely information to make better decisions." One percent may sound small, but considering the millions spent annually on the TB management program, it equates to many thousands in annual savings. Further, what is just the beginning for New Zealand's AHB may very well be the beginning of the end for bovine TB in the island farming nation; a savings that is incalculable.

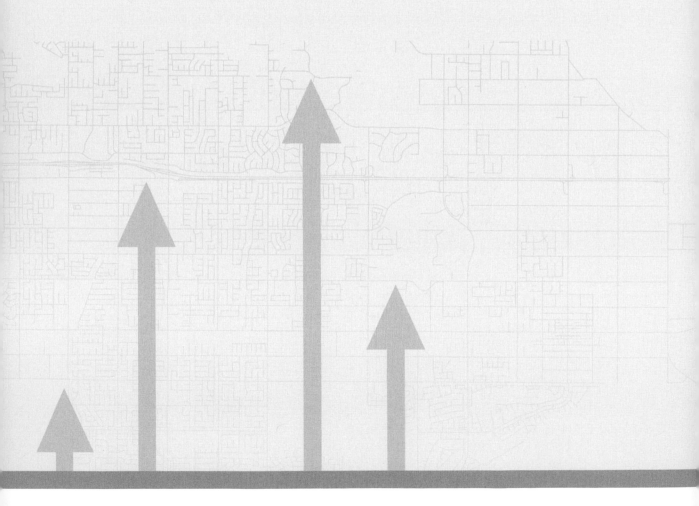

Chapter 3

Avoid costs

Governments face increasing demands for services even when their budgets stay flat. Today's citizens are accustomed to plucking the information and services they require directly from the web, and they expect the same type of on-demand services from their public institutions. Agencies are meeting these demands every day without increasing their budgets by using GIS to help avoid costs.

Cost avoidance is sometimes referred to as "soft savings," but the impact is real. It occurs in one of several ways, but in each case new expenditures are avoided because system capacity is sufficient to handle new demands without additional staff or tools. In many cases, cost avoidance occurs as a by-product from well-constructed workflows assisted by GIS.

Numerous examples of GIS-derived cost avoidance exist. Web-based GIS solutions provide on-demand access to data and information that help avoid customer service costs, especially when current events spike the public's demand for information. GIS improves the management of public assets such as roads and utility infrastructure, leading to better maintenance routines that, in turn, help avoid costly repairs. Governments use mobile GIS technologies for a variety of needs, such as enabling work crews to upload data to the office directly from the field to avoid travel time and data postprocessing. Another is the geographic intelligence provided by GIS to guide vehicle fleets as they provide such public services as waste removal. By knowing in real time the location of every truck, managers can optimally reroute crews as needed without incurring additional costs.

The calculation of cost avoidance goes deeper than simply comparing the before and after impacts of GIS implementation. Instead, it's about showing how GIS prevents new costs from ever arising. With a little research, most will find numerous examples of GIS-derived cost avoidance.

Baltimore County, Maryland, strategic plans reveal millions in return on investment from GIS

SECTOR: Information technology (IT)
Rob Stradling, Chief Information Officer, Baltimore County, Maryland

Baltimore County, located in northern Maryland, has used GIS since the mid-1980s when it invested in its first license of Esri's ARC/INFO software. The size of the system and its uses have grown considerably over time. Today, the county's GIS is an enterprise solution available to all of its agencies and offices. The county also gives the public access to its GIS maps, data, and services through a fee-based program. Organizations or individuals can receive printed copies of published maps or gain access to the digital data and services, allowing them to create custom mapping products based on their own specifications.

Along with the growth of Baltimore County's GIS and the services it provides to staff and the public, there have also been increases in the cost to operate and maintain the system. As a result, the county's Office of Information Technology (OIT), along with the Office of Budget and Finance, initiated a project to develop a strategic business plan for their GIS.

Charting the future course of GIS in the county required a solid understanding about how the system was currently being used and, importantly, determining what the county was getting back from its investment in the technology. Considering this, a significant portion of the planning work involved a return on investment (ROI) study to find out how the county's GIS measured up when its benefits were compared to its costs. From the ROI study conducted as part of its strategic plan, Baltimore County uncovered hundreds of workflows assisted by GIS and determined that its total investment in GIS was returning more than $4.5 million in gross annual benefits.

> From the ROI study conducted as part of its strategic plan, Baltimore County uncovered hundreds of workflows assisted by GIS and determined that its total investment in GIS was returning more than $4.5 million in gross annual benefits.

An inventory of GIS applications

The county partnered with Dewberry, a nationwide professional services firm, to augment its planning team and help conduct the study. Rather than examining the historical costs and benefits of the county's GIS technology, the group decided to concentrate on the current state of the county's GIS.

The planning team began by examining how GIS was being used across the county's departments and agencies. The team performed a detailed inventory of all the GIS data layers used and maintained by the county, then researched who within the county was using GIS and for what purposes.

The list of GIS users was then used to set up a series of interviews in which the planning team asked questions to draw out those business processes assisted by GIS and GIS

All told, the inventory and interview process revealed several hundred workflows that used GIS in the county.

data. In addition to the interviews, several other methods to gather information were used, such as short- and long-form questionnaires, and follow-up phone and face-to-face interviews to compile data about the GIS infrastructure, comparable industry practices, and public access programs within the county.

The information gathered during the interviews was compiled into separate reports that summarized GIS usage within each county agency. These reports were then presented back to agency stakeholders through a series of workshops that were designed to verify and build consensus around the findings.

All told, the inventory and interview process revealed several hundred workflows that used GIS in the county. With the overall use of GIS fully uncovered and documented by the planning team, the focus of work shifted to establishing the costs and the benefits associated with Baltimore County's GIS.

A remarkable return on investment

The planning team's method for measuring cost was based on a simple principle: any expenditure required to support GIS activities within the county was considered a cost. This included money spent for personnel salaries, database maintenance and administration, hardware, software, staff development, and capital expenditures. The cost analysis also included expenditures for miscellaneous supplies and administrative support. To help organize its findings, the planning team developed a categorization system that broke costs down by whether they were incurred to support enterprise (countywide) uses, agency (departmental) needs, or standard capital expenses.

To establish benefits, the planning team examined each workflow in its inventory to determine specifically how GIS was being used to assist it. These cases ranged from using GIS to help determine the location of sidewalk ramps to protecting and managing groundwater resources to maintaining an inventory of all county-owned bridges. For each of these workflows, the team then examined how much time it would take to complete these activities with and without GIS. Once completed, the results of this work became the primary basis for the cost-benefit analysis.

By analyzing and comparing the time spent to perform an activity with and without GIS, the planning team derived a time-savings benefit. The total hours saved were then multiplied by a standard rate of $33.95 per hour, and that cost figure then became the personnel savings

Activity: Install Roundabouts					
Annual Time Savings from the Use of GIS					
Staff Hours w/o GIS (Manual)	Staff Hours with GIS	Difference	Annual # Iterations	Total Hours Saved Using GIS	Annual Time Savings Benefit (Based on $33.95/hr)
70	21	49	2	98	$3,327.10

For each workflow aided by GIS, the planning team examined how long it took to perform the task with and without GIS. The results of this work allowed the team to objectively determine and report how much time was saved using GIS. The table shows the results of this analysis for a public works task related to the design and construction of traffic roundabouts. Courtesy of Baltimore County, Maryland.

calculation. Also included in the benefits tabulation were the costs avoided by using in-house GIS capabilities (rather than paid consultants) to assist with various project requirements and the revenues generated by licensing GIS data and services to the public and businesses through the county's website.

> # GIS delivered a net benefit of just over $2.5 million a year by reducing the time it takes staff to carry out daily business operations.

With the full set of benefits and costs established across all of the county's workflows, the planning team summarized the results by agency. The results showed that nearly all agencies within the county were receiving a considerable

ROI from their use of GIS, and that for the county as a whole, GIS delivered a net benefit of just over $2.5 million a year by reducing the time it takes staff to carry out daily business operations.

A clear business case for GIS

At the end of the planning process, Baltimore County received a strategic business plan that documented the full ROI coming out of its current GIS along with a direction forward for future uses and workflow integrations. The results showed that the county's GIS was extremely viable and its returns so significant that one could hardly argue against further centralization and use of GIS within the county.

Baltimore County's senior managers, like most senior managers, are fiscally minded. Approvals for new technologies or realignments of business operations only come if there is a clear business case associated with them. Rob Stradling, Baltimore County's Chief Information Officer, highlighted this point while discussing the importance of the ROI finding in the county's strategic plan when he said, "Bottom line is the return on investment."

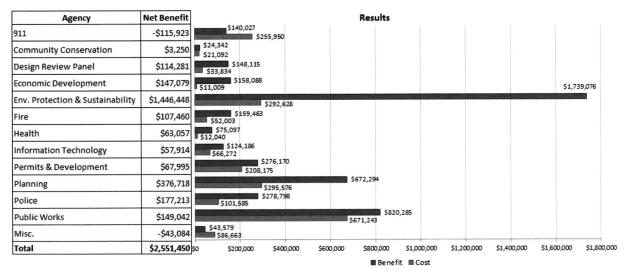

Agency	Net Benefit
911	-$115,923
Community Conservation	$3,250
Design Review Panel	$114,281
Economic Development	$147,079
Env. Protection & Sustainability	$1,446,448
Fire	$107,460
Health	$63,057
Information Technology	$57,914
Permits & Development	$67,995
Planning	$376,718
Police	$177,213
Public Works	$149,042
Misc.	-$43,084
Total	$2,551,450

This chart summarizes the net benefits derived from GIS for the county agencies that use it. The horizontal bars shown for each agency represent benefits (red) and costs (blue). As shown here, the planning team determined that, for the county as a whole, GIS delivered a net benefit of just over $2.5 million a year. Courtesy of Baltimore County, Maryland.

Baltimore County's GIS Benefits
119,377 Hours saved through the use of GIS.
$4,052,895 Money saved through the use of GIS.
$606,626 Costs avoided from the use of GIS.
$4,659,521 Total annual benefits realized through the use of GIS.
121% Percentage of annual benefits realized through the use of GIS after its cost is subtracted.
221% Percentage of money saved through the use of GIS.

ROI Summary of Baltimore County's GIS. Courtesy of Baltimore County, Maryland.

With the backing of the county's senior management, the plan became the starting point for a more efficient approach for managing GIS within the county. GIS personnel were reassigned to streamline business operations, and the OIT developed formal GIS Service Level Agreements that clearly defined the terms for integrating GIS infrastructure, products, and services into agency operations. Another benefit was that it increased the communication between agencies about GIS activities and database needs. Overall, the study charted a course forward for the use of GIS and presented an opportunity for the agencies to further integrate GIS into their business processes using a centralized approach that served the entire enterprise without redundancies.

The strategic planning work and its associated cost-benefit analysis brought clarity to the fact that GIS is an integrated part of the county's daily operations, one that saves several million dollars a year through more efficient workflows and revenues generated. What began as an exercise to establish whether GIS was producing tangible results led to a clear realization that GIS was providing a critical service to the county on a daily basis, and that even with all the benefits and savings it was producing, there were still plenty of new opportunities for GIS to improve workflows and help the county avoid costs—opportunities that the county are now making a reality.

> The strategic planning work and its associated cost-benefit analysis brought clarity to the fact that GIS is an integrated part of the county's daily operations, one that saves several million dollars a year through more efficient workflows and revenues generated.

Bonner County, Idaho, manages invasive weeds with GIS

SECTOR: Environmental management
Terry McNabb, AquaTechnex

Lake Pend Oreille is an important ecological and economic resource in Bonner County—located near the northern tip of Idaho—that faces a growing threat from the aquatic weed Eurasian milfoil. This invasive species rapidly replaces the native aquatic vegetation while degrading water quality and natural habitats. It also hampers recreation by clogging local marinas and beaches with dense weed mats.

To cut back its growing problem, the county's Public Works Department hired an aquatic weed control expert, AquaTechnex, to survey the extent of the infestation and develop a comprehensive treatment plan. In an effort to keep

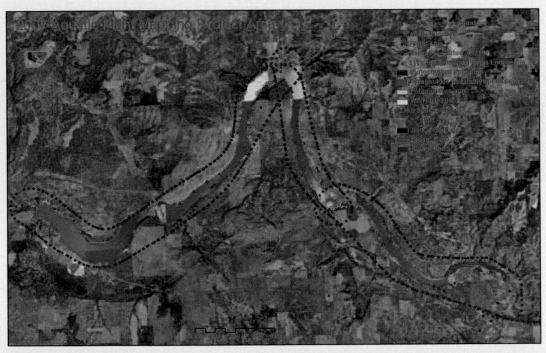

This project map shows the flight lines created during an aerial survey. The green dots represent the location of an aerial image collected at that exact data point; during analysis, biologists can click each green dot to view and get details about the image in a separate window. Courtesy of Terry McNabb, AquaTechnex.

the budget under control, the county went with a relatively low-cost approach that combined GIS, GPS, and mobile technologies into a successful solution.

The weed abatement team began with an aerial survey, documenting the full extent of the infestation along nearly one hundred miles of shoreline. Rather than using traditional aerial technologies, which come with a very high price tag, the team established an accurate basemap of imagery by capturing a succession of photos from the plane using a handheld digital camera connected to a Trimble GPS device.

Back on the ground, technicians used ArcMap to tie each image to its position along the flight line and map the weed beds visible on the imagery within a new GIS layer. This layer was then used to guide boat trips to infested areas where crews refined the plume boundaries and measured their depth.

The use of mobile GIS and GPS technologies along with desktop mapping helped Bonner County avoid over $50,000 in project costs. The survey and mapping were completed in only five days and for $6,000. Had the contractor used conventional methods to collect the imagery and run the field work, the price tag would have risen to nearly $60,000.

> # The use of mobile GIS and GPS technologies along with desktop mapping helped Bonner County avoid over $50,000 in project costs.

Bonner County used the survey results to secure $1.8 million in state grant funds to execute its comprehensive treatment plan. During the treatment work, the spatial data created from the survey was used to plan for and carry out the precise application of herbicides, saving both labor time and material costs. One year after the initial weed management plan began, an independent expert from Mississippi State University confirmed that the treated areas experienced "very good to excellent control."

Los Angeles Bureau of Sanitation uses GIS to avoid costs and improve workflows

SECTOR: Solid waste

Salvador Aguilar Jr., Environmental Engineering Associate, Bureau of Sanitation, City of Los Angeles, California

The critical job of managing solid waste removal for the citizens of Los Angeles falls to the city's Bureau of Sanitation. The second most populated city in the United States produces about 1.5 million tons of solid waste a year (about 0.7 million recycled), and at any given moment within the city, there are 680 to 720 trucks running their daily routes to pick up and haul trash and other types of waste to their appropriate destinations. It is a complex system, one in which GIS and GIS-based logistics play a very important role in making it all work as efficiently as possible.

While talking about how GIS is used to assist in this work, Salvador Aguilar Jr., an environmental engineering associate at the Bureau of Sanitation, stated, "We have a lot of operations that without GIS would be very labor intensive." He added, "Bottom line is that it allows us to do more work with less people." For example, when it came to systematically replacing all trash containers in the city, the analytical assistance provided by GIS allowed the city to avoid an estimated $400,000 per year in salary costs.

The analytical assistance provided by GIS allowed the city to avoid an estimated $400,000 per year in salary costs.

One of the primary ways the Bureau of Sanitation uses GIS is to develop routing solutions. For example, it uses Esri's ArcLogistics to create daily point-to-point driving routes for the approximately 50,000 special collection requests the city receives each month. RouteSmart, an Esri partner solution, is used to generate and maintain the continual routing system used to carry out Los Angeles's standard curbside services, which haul off trash, recyclables, and yard waste each week from roughly 2.3 million curbside containers spread across the city.

The Bureau of Sanitation relies heavily on the routing capabilities of GIS to pick up trash from roughly 2.3 million curbside containers. Courtesy of City of Los Angeles, California.

The Bureau of Sanitation's GIS staff is developing a web-based application using ArcGIS for Server to analyze the efficiency of its hauling routes. The web-based GIS application lets the bureau see and examine the distribution of workloads, tonnage carried, dump site break-points, overtime incurred, system delays, and miles driven. This information, when viewed in a spatial context, raises the situational awareness of the overall system and is used to continually fine-tune routing and dispatch operations, making the work more efficient over time.

Bottom line is that it allows us to do more work with less people. Salvador Aguilar Jr., Environmental Engineering Associate, Los Angeles Bureau of Sanitation

Chapter 4

Increase accuracy

Public policy requires accurate information to make informed decisions. In fact, most projects begin with identifying the best and most reliable information sources. In today's era of transparency and accountability, citizens demand a window into the methodology governments use to make decisions, and they want to see the information in a geographic context to which they can relate.

The Internet has exposed hundreds of data sources on a wide array of topics. GIS professionals increase the confidence end users have in the data by building it on standardized data models that establish authoritative geospatial information.

Today's GIS professionals construct their spatial data from a wide variety of sources using a collection of methods that provide the most accurate and up-to-date facts to the most departments or individuals. The basemaps used to build the foundation of the GIS are generally derived from an original source such as sur- veying or engineering drawings. The system will cleanse or verify the accuracy of information by searching the database and isolating questionable data inputs. Several different departments might input address data as a key field; by cross-referencing these data sources and defining data standards to create a more accurate data- base, GIS delivers a more accurate result. Moreover, the integration of precision support technologies, such as GPS, orthophotography, remote sensing, geopositioned video, and satellite imagery, increases the validity of GIS databases. Data derived from citizen sensors using smart devices provides up-to-the-minute changes in the reporting of attribute characteristics, such as whether a water main is broken or street light is out.

Accurate information contributes to the improvement of an organization's effectiveness and benefits many departments within a business or agency. In planning or public works departments, an accurate GIS could affect a decision to include or exclude a group from a program or fee structure. For instance, the number of homeowners who purchase additional insurance to comply with revised flood insurance rate maps could change dramatically when GIS combines multiple databases for enhanced accuracy.

Accuracy also helps improve efficient and productive performance. While GIS is a technology and science designed to apply geographic analyses and reporting to the world's problems, the results of these functions are highly dependent on accurate spatial and tabular data sources. GIS increases accuracy, and that results in better decisions, analyses, reports, routing solutions, and other products.

Charting the roads that connect the vast Navajo Nation

SECTOR: Transportation

Jonah Begay, GIS Supervisor, Navajo Division of Transportation

Nick Hutton, Director of Asset Management, Data Transfer Solutions, LLC

Spanning approximately 27,000 square miles across Arizona, Utah, and New Mexico, the Navajo Nation is the largest sovereign nation in the contiguous United States. It has a strong presence in US government and often leads the way in tribal efforts to promote key areas such as economic development, health care, and education at the national level. Despite its prominence, the sheer size and remote nature of the Navajo Nation presents unique challenges in managing its infrastructure and resources.

Consider, for instance, the road inventory that tribes submit each year to the Indian Reservation Roads (IRR) program. The IRR program maintains the official inventory of reservation roads in the United States and is designed to allocate federal funding to tribal governments for transportation planning and road maintenance activities. To ensure that all of its roads are accurately accounted within the IRR inventory, the Navajo Division of Transportation (DOT) developed an integrative solution that uses GIS as its base technology. It's a solution that nearly doubled the roads accounted for within DOT's inventory, increased IRR funding by 30 percent, and produced a fifteen-fold return on the initial project investment.

A component of the broader Integrated Transportation Information Management System (ITIMS) program, the Bureau of Indian Affairs (BIA) Division of Transportation maintains the national reservation road inventory in a system called RIFDS (Road Information Field Data System). Each year, as part of the IRR program, tribes are eligible to submit their road inventory data to one of the twelve BIA regional offices. There are approximately 560 nationally recognized tribes that fall under the twelve BIA regions. The Navajo Nation submits its road inventory to the BIA Navajo Regional Office (BIA–NRO) in Gallup, New Mexico, and the accuracy and thoroughness of its annual submission plays directly into the amount of resources the tribe receives for its road maintenance programs.

The Navajo road inventory was far from comprehensive. In early 2006, its official RIFDS inventory contained approximately 9,800 miles of roads. Roughly 6,000 miles were BIA roads, and the remaining 3,800 were primarily state and county roads, with very few tribal roads mixed in. Navajo transportation officials determined that the road inventory was substantially underperforming in the following two key areas:

- Road mileage quantity—The current inventory reflected only a small percentage of the reservation's tribal roads. It was widely believed that there were thousands of miles of tribal public roads that were eligible for the inventory, but were not yet included.
- Data quality—Of the 9,800 miles of roads in the 2006 inventory, only a portion generated funding in the RIFDS allocation formula. Some roads in the existing inventory were missing key pieces of information that excluded them from funding. Misinterpretations of program regulations resulted in a lack of quality data, exacerbating the effect of the low mileage numbers.

To address the omissions and inaccuracies in the road inventory, the Navajo DOT launched a proactive and aggressive campaign that would expand its internal capacities, establish a systematic method for identifying eligible

The sheer size and remote nature of the Navajo Nation presents unique challenges for managing and maintaining its roads. To assist in this work, the Navajo Division of Transportation uses GIS to help maintain a digital and map-based roads inventory. Courtesy of Navajo Division of Transportation.

public tribal reservation roads, remove subjectivity from regulations, and build a GIS-based system to improve both the quantity and quality of the road inventory data. With the support of the Navajo Nation Transportation and Community Development Committee (TCDC) and under the direction of former Navajo DOT Director Tom Platero, the Navajo Inventory Team and consulting project manager Nick Hutton embarked on an innovative and challenging endeavor that would span more than four years.

The first step was to fortify the Navajo DOT's existing technology infrastructure. New enterprise-class servers were put into place, network bandwidth was expanded, and new data was collected. The Navajo DOT implemented a GIS-enabled, multitiered, web-based information architecture that was part of an integrated hardware and software solution provided by the InLine Corporation (now IceWeb) and Esri. IceWeb servers were preloaded with ArcGIS for Server–Enterprise Edition and Microsoft SQL Server and were preconfigured to optimize system performance. This saved many hours of work by allowing the Navajo project team to focus on developing core programs and data instead of testing and tweaking the new system.

The next step was to obtain and develop the required GIS data. The project team was able to acquire current, reservation-wide aerial photography captured as part of a joint project between the US Department of the Interior and the State of New Mexico. Once the imagery was loaded onto the new system, it was time to start digitizing the GIS layer of road centerlines.

Along with a team of GIS technicians using Esri's ArcGIS technology, GIS consultant and Esri partner Data Transfer Solutions (DTS) began the digitization process. It was not until this time that the team realized the full extent of the project. After several months of digitizing, the team mapped more than 70,000 miles of roads and trails. While not all of the digitized roads were eligible for the official IRR inventory, the potential challenges associated with managing these roads were daunting to Navajo DOT officials. This realization underscored the notion that the automation of data entry and mapping provided by GIS would be an absolute necessity in the development of the Navajo DOT road inventory system. While the GIS techs continued the digitization process, the programming staff at DTS and the Navajo DOT project team were busy developing the inventory management system from the GIS technologies.

The team concluded that the system must be secure, web-based, GIS-enabled, usable by staff members both with and without technical expertise, and capable of mapping automation—specifically, strip-map automation to rapidly generate standardized maps along the full extent of selected road corridors. In addition, the team identified the need for a robust querying component that included bidirectional filtering that allowed users to select roads in the map display to access their associated table records, or inversely, make tabular selections to locate the associated road features on the map.

What emerged was a system the Navajo DOT calls NAVRIS (Navajo Roadway Inventory System). At the heart of the system is a GIS, and in addition to its web and automation capabilities, NAVRIS incorporates a series of validation scripts to ensure the data is entered in accordance with program requirements.

One of the most challenging aspects of the project was establishing consistent interpretations of the IRR

This realization underscored the notion that the automation of data entry and mapping provided by GIS would be an absolute necessity in the development of the Navajo DOT road inventory system.

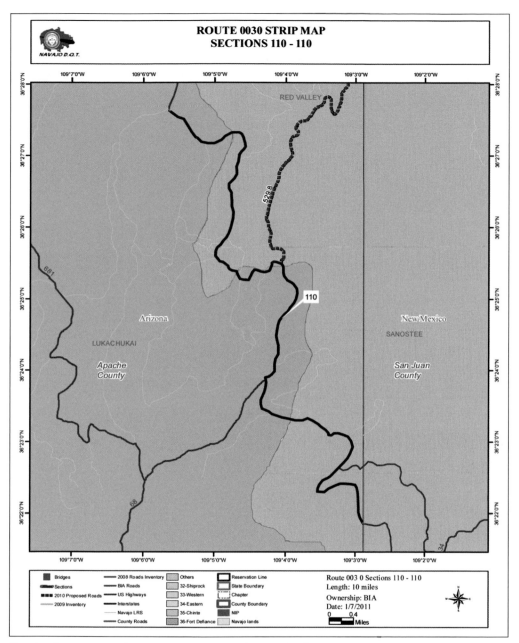

The development team leveraged GIS to create functionality for the Navajo DOT that automated the production of standardized map products like the one shown here. This functionality permits the rapid production of maps with a consistent design along any chosen road corridor. Courtesy of Navajo Division of Transportation.

The GIS allows NAVRIS users to seamlessly interact with tabular and spatial views of the roads inventory. For example, users can select tabular filter (query) results to view their associated road features in the map display, or invert the process by selecting road features on a map to access their associated tabular data. This type of functionality, which is unique to GIS, is used by technicians to efficiently review and update the road data within the inventory. Courtesy of Navajo Division of Transportation.

> **As a result of the GIS functionality and automation tools within NAVRIS, the percentage of roads questioned by the BIA because of missing or incorrect data has declined dramatically.**

a systematic approach to maintaining its road inventory. Beyond supporting the immediate needs of the federal IRR program, NAVRIS serves as the foundation for a comprehensive infrastructure management system to support Division of Transportation activities.

Today, the Navajo DOT continues to leverage the GIS functionality within NAVRIS as part of its ongoing IT strategy. NAVRIS offers the first consistent, verified interpretations of IRR regulations and the ability to programmatically generate the required BIA deliverables. By taking the initiative to build a geospatial road inventory program that helps define and facilitate the IRR process, the Navajo DOT has become a stronger, more sophisticated tribal entity with more time and resources to support the development and maintenance of its expansive infrastructure.

program regulations between the BIA–NRO and the Navajo DOT staff. This took many months of research in collaboration with BIA–NRO Chief Engineer Harold Riley and his staff. To the credit of both agencies, considerable common ground was established, and the findings were subsequently programmed into the core automation and validation logic of the system. As a result of the GIS functionality and automation tools within NAVRIS, the percentage of roads questioned by the BIA because of missing or incorrect data has declined dramatically.

As of the 2010 IRR submission cycle, the Navajo DOT significantly increased the number of miles in its inventory. It grew from 9,800 miles in 2006 to nearly 16,000 miles, including approximately 6,000 miles of tribal roads. The improved accuracy and additional mileage added to the existing roads data increased the Navajo Nation's IRR funding by an average of 30 percent compared to its funding level prior to implementing the GIS-based solution. To date, the Navajo Region has received a fifteen-fold return on the Navajo DOT's initial investment in the IRR project. This adjusted allocation will allow for critical transportation infrastructure improvements supporting access to education, employment, health care, and other services for the Navajo Nation's widespread residents.

By creating NAVRIS and several supplemental data policies and standards, the Navajo DOT has developed

> **The improved accuracy and additional mileage added to the existing roads data increased the Navajo Nation's IRR funding by an average of 30 percent compared to its funding level prior to implementing the GIS-based solution. To date, the Navajo Region has received a fifteen-fold return on the Navajo DOT's initial investment in the IRR project.**

City of Alpharetta, Georgia, uses GIS to get an accurate census count

SECTOR: Planning

Steven Wardrup, GISP, GIS Manager, City of Alpharetta, Georgia

Adam Montgomery, Systems Manager, City of Alpharetta, Georgia

The city of Alpharetta, Georgia, is located twenty-five miles north of Atlanta. The city's residential population is approximately 50,000, but on weekdays its daytime population nearly doubles from the employees who commute into town to work for one of the several large companies located within the city.

The daily influx of population into Alpharetta places a high demand on all city functions, from emergency services to public works. As with most US cities, Alpharetta receives some federal and state funding to support these services. This funding is often directly proportional to its current census figures. Considering the importance of a correct census count, Alpharetta used its GIS to catch and correct nearly two thousand addresses missing from the US Census Bureau's records, a critical fix made possible through GIS, and a fix that equated to a nearly 10 percent increase in the city's reported population.

> Alpharetta used its GIS to catch and correct nearly two thousand addresses missing from the US Census Bureau's records, a critical fix made possible through GIS, and a fix that equated to a nearly 10 percent increase in the city's reported population.

The city uses GIS on a daily basis to assist with workflows within its seven departments. These uses range from improving 9-1-1 call plotting to managing city assets such as traffic signals and road signs. Alpharetta also maintains a current stock of base data that includes imagery, NAVTEQ street centerlines, and boundaries. Beyond assisting of day-to-day workflows, the city's GIS provides a ready-made solution to various projects as they arise. This was the case when Alpharetta needed to respond to the Census Bureau's Local Update to Census Addresses (LUCA) program.

The LUCA program essentially gives cities and counties the opportunity to review the Census Bureau's mailing list for their jurisdiction. This list is the basis for sending out census questionnaires, so ensuring its accuracy is critical to getting a complete census count within any jurisdiction.

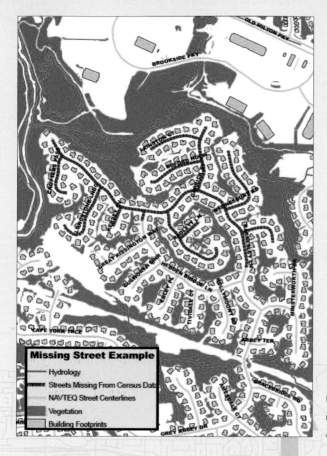

Missing Street Example
— Hydrology
— Streets Missing From Census Data
— NAVTEQ Street Centerlines
■ Vegetation
□ Building Footprints

Using GIS, Alpharetta quickly pinpointed addresses missing in the Census Bureau's records. Courtesy of City of Alpharetta, Georgia.

Alpharetta used GIS to map the address list provided by the Census Bureau and compare it with the current data. GIS made an otherwise laborious and time-consuming process straightforward and simple. Without GIS, Alpharetta would have had to manually compare hard-copy maps and address lists. The process enabled Alpharetta to locate 1,937 addresses missing from the Census Bureau's records. Using the accepted standard of 2.5 persons per household, the correction equated to an increase of nearly 5,000 persons, or a roughly 10 percent increase in the city's total population.

> Alpharetta used GIS to map the address list provided by the Census Bureau and compare it with the current data. GIS made an otherwise laborious and time-consuming process straightforward and simple.

Hudson, Ohio, increases government transparency and workflow accuracy using GIS

SECTOR: Public works
Paul Leedham, GIS Manager and Database Administrator, City of Hudson, Ohio

The city of Hudson is located in northeast Ohio between Cleveland and Akron. It is a relatively small community with a population of just over 22,000. Like many towns its size, Hudson's residents and city officials keep a close watch on how their local tax dollars are used.

The city's Information Technology (IT) Department built its GIS incrementally beginning with a web-based approach that used Esri's Internet mapping technologies to share GIS data and maps. From this unique starting point, IT developed the GIS into an enterprise solution that now serves all of the city's departments and the public as well.

As the GIS grew, so did the number of questions about its costs and purpose. To help bring clarity to these questions, Hudson's GIS Manager Paul Leedham formally defined how the city was using GIS and the return on investment it was getting from GIS. By doing this, Leedham was able to document several positive returns. For example, he found that GIS saves Hudson just over $90,000 annually by increasing the initial accuracy of engineering and design projects and the precision with which field crews can locate underground utilities prior to construction work.

> ## GIS saves Hudson just over $90,000 annually by increasing the initial accuracy of engineering and design projects and the precision with which field crews can locate underground utilities prior to construction work.

Hudson's GIS staff has been able to leverage its initial success with GIS to develop the system further. Staff members have created several online GIS applications to assist workflows, provide more government transparency, and create feedback loops that increase the accuracy of the underlying spatial data. One example of this is a web-based GIS solution used by the city to manage and track the status of the work orders its departments are responding to on a daily basis. The application is also available to the public, which can use a map-based interface to get accurate information about where city crews are working across the entire city.

Speaking about the work order tracker and other public-facing GIS solutions used by the city, Leedham said, "This is the kind of thing that lends a greater transparency into government services." By using GIS to share information with

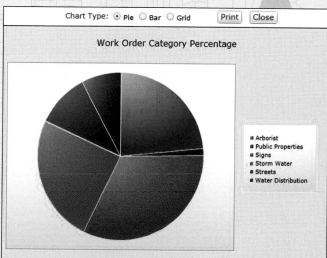

The work order tracker of the City of Hudson. This GIS solution improves the transparency of Hudson's daily activities by allowing users to query the status of work orders and view them on the map. Users can even generate reports that summarize the amount and type of work going on in a particular area.

Courtesy of City of Hudson, Ohio.

city crews and the public, the city has created feedback loops that catch errors in its data that might otherwise have gone undetected. The cycle improves the accuracy of its spatial data, which in turn improves city functions through more reliable GIS services.

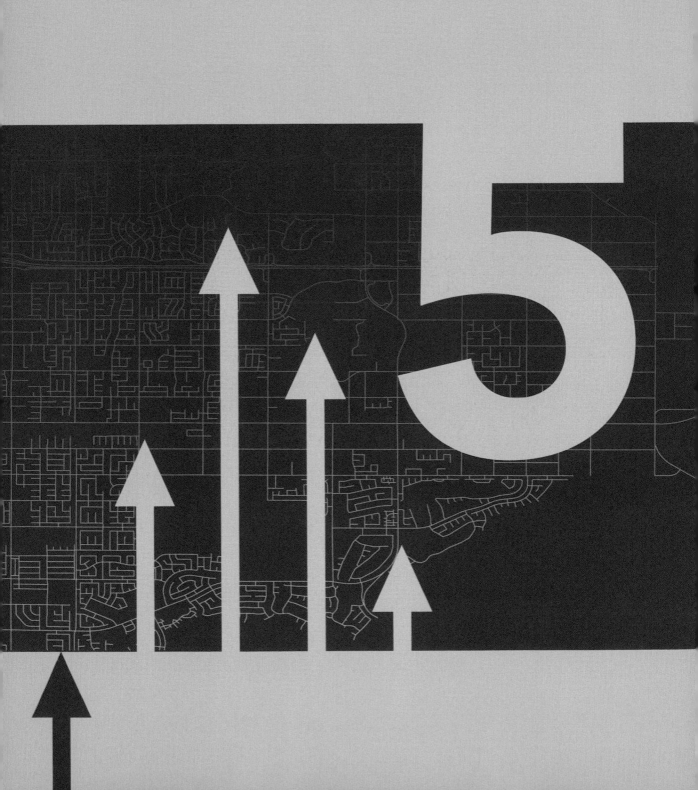

Chapter 5

Increase productivity

The ratio that measures how well governments convert inputted resources—labor, materials, machines—into goods and services is called productivity. From a service level perspective, improvements in productivity result from a measurable increase in the output of product or service levels. Labor productivity refers to the relationship between output and the labor time used to generate that output. Increases in productivity are typically reported as a percentage increase in yield or via a ratio of output per hour.

Government organizations adopted new ways to increase productivity to meet accountability and performance demands from the public. The public sector views improved productivity as critical to the management of time, labor, and costs. Management constantly weighs new approaches to increase productivity. These methods have evolved to include solutions such as employee motivation, improved equipment and technology, and ergonomics.

As labor and economic resources dwindle and older procedures offer diminishing returns, organizations seek new opportunities to improve performance. Just as managers view GIS as a tool to increase efficiency in reengineering processes, they are looking at the roles geography and GIS play within daily operations.

Geography has traditionally played a role in prioritizing workloads as a means of meeting service schedules. GIS allows a migration from static service boundaries such as code enforcement districts, inspection territories, and police beats as the devices for load balancing to dynamic resource allocation and on-demand scheduling. Compelling evidence shows that GIS has transformed the way we work.

Within the public sector, GIS technology is a proven and powerful tool for improving workflows and broadening the scope of new initiatives like Government 2.0. Mobile GIS solutions have also increased the performance rates of public works professionals in data collection and asset management by incorporating spatial data into work order management systems that balance both existing and on-demand workloads with respect to location of service demands and resource availability.

GIS-based asset management peaks productivity for Colorado Springs

SECTOR: Public works
Andy Richter, Traffic Technician II, City of Colorado Springs, Colorado

In 2006, *Money* magazine ranked the City of Colorado Springs, Colorado, "the best place to live" in the category of cities with 300,000 or more people. With this popularity, however, come growth and added pressure on city infrastructure.

Due to the fast pace of development, Colorado Springs began having a difficult time managing the growing number of requests coming into the city for the repair and installation of street signs. The total number of calls exceeded 10,000 per year, and the Public Works Department staff was struggling to keep pace. The city's lead for asset management work in the Traffic Division, Andy Richter,

> The problem with all these requests and work orders was that we had no information about asset inventory, and we were getting buried in paperwork. Andy Richter, Traffic Technician II, City of Colorado Springs, Colorado

summarized two key aspects of the challenge by saying, "The problem with all these requests and work orders was that we had no information about asset inventory, and we were getting buried in paperwork."

The traditional work order process for asset management at the city was paper based, very time-consuming, and disconnected. Requests for sign repairs or new installations were reviewed and converted to work orders in one building, sent by interoffice mail or the postal system to another building for prioritization, then delivered to a third location for assignment to work crews. After a crew finished each job, the paperwork was mailed back to the building where it was originally received and logged as completed.

The siloed and paper-based system created a paper trail that made it difficult to track the status of work orders as they were passed from one group to the next. This lack of awareness about the overall status of work often led to duplicate work orders for the same incident, which was hindering productivity in the field. The system was too slow and needed improvement; a typical work order for a sign repair was taking nearly fifty hours from start to finish, costing the city approximately 30,000 hours of staff time annually.

A spatial solution

The growing challenges caused by the outdated work order system were made clear to city officials who, in response to the problem, approved the creation of an asset management

> # Their study made it clear that, regardless of what product or methodology they chose for asset management, GIS would be the foundation of the solution.

team. Team members researched best practices for asset management, specifically looking into applications for sign maintenance. They reviewed proposals from vendors, met with public works departments in cities around the state, and, throughout the work, kept city staff and management with a stake in the project involved. Their study made it clear that, regardless of what product or methodology they chose for asset management, GIS would be the foundation of the solution.

After completing product evaluations, the city chose to work with Cartegraph, an Esri partner. Cartegraph provided an asset management solution that ran directly on ArcGIS for Desktop and included a suite of GIS-based applications that the team could use to assist all aspects of asset management, from managing work order requests at call centers to managing crews working in the field. The solution also permitted them to create customized forms that were similar in form—but digital—to the ones their staff had used in the past. Shortly after launching the new software, work crews were also equipped with GPS and ruggedized laptops running ArcGIS for Desktop and the Cartegraph software.

Breaking down silos and increasing productivity

The switch to a GIS-based approach for asset management had an immediate positive impact on all the workflows. GIS added locational awareness to each part of the process, removed the sluggish back and forth of paperwork, and connected the flow of information from one phase of the work to the next.

Under the old analog system, the location of a requested work order was manually marked on a paper map after being submitted by customer service staff. With the new system, the location was automatically displayed alongside all of the other active work orders on an interactive GIS map immediately after it was entered into the system by a call taker. Considering that a damaged street sign can trigger multiple calls from the public, this on-the-fly mapping allowed call takers to quickly determine whether the event already has a work order assigned to it, avoiding the possibility of creating a duplicate ticket for the event.

The biggest gains in productivity have occurred in the field. The mobile GIS kits carried by the work crews enable them to see the spatial distribution of work orders and, while in the field, plan their routes from one incident to the next. This logistical improvement has saved the time it takes crews to get from one job to the next, which, in turn, saves fuel costs. And, because crews log the data associated with each incident while on site using customized digital forms, they do not need to fill out paperwork when they get back to the office at the end of the day.

The real-time situational awareness and connected workflows provided by the system have made the supervision of work more effective. Managers use the solution to track job progress from start to finish and to see where crews are currently working. The spatial perspective provided by GIS makes it possible to efficiently respond to unexpected work orders with a high priority—such as a

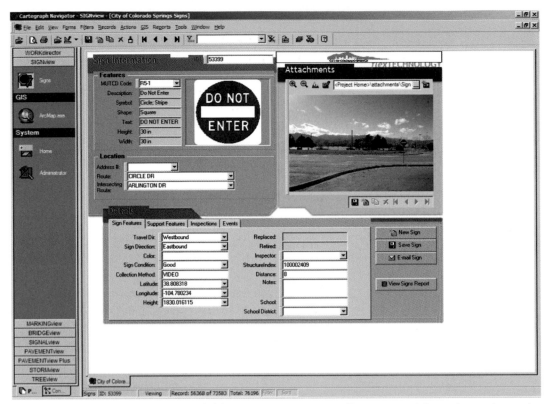

After clicking a sign in the locator map, field crews can see a picture of it, enter new information, and amend incorrect details. Approximately 2,000 work orders a year were being completed just before the new system was installed. Today, through the gains provided by the GIS, the number has jumped to over 13,000 and without additional staff. Courtesy of Andy Richter, Traffic Technician II, City of Colorado Springs, Colorado.

stop sign that has fallen—because managers can see which crews are already in the vicinity and redirect them as needed. Once the tasks associated with a work order are completed, the supervisor is automatically notified that the job is done, and the full history of the record is then archived within the system.

All signs reflect success

The city's investment in GIS for asset management led to dramatic increases in productivity. Speaking about this impact, Richter said, "Going electronic, we are able to eliminate five steps of our workflow—most of it related to paper processing—which decreases the time it takes to complete a work order from forty-nine hours to eighteen minutes. Increased productivity, without having to add new staff, gives us a huge return on our investment."

The drastic decrease in the time it takes to respond to and complete a work order means more work gets done every day, and the increase is striking. Approximately 2,000 work orders a year were completed just before the new system was installed. Today, through the gains provided by the GIS, the number has jumped to over 13,000 and without additional staff.

> Going electronic, we are able to eliminate five steps of our workflow—most of it related to paper processing—which decreases the time it takes to complete a work order from forty-nine hours to eighteen minutes. Increased productivity, without having to add new staff, gives us a huge return on our investment. Andy Richter, Traffic Technician II, City of Colorado Springs, Colorado

The increases in productivity freed time for other important projects, such as the completion of a full sign inventory. Using their mobile GIS capabilities in combination with the Cartegraph software, crews inspected and cataloged 73,000 street signs—something they could not do before they added GIS to the mix. With the completion of the sign inventory, the two primary needs noted by Richter had been fulfilled; the city now had a paperless work order system that was extremely efficient and a complete inventory of their road signs.

The benefits of the new system and its derivative effects speak for themselves. Describing an encounter with a city administrator, Richter said, "The assistant city manager asked me how many 25 mph signs we had. In less than five minutes, not only could I tell him, but I could show him on a map." Richter pointed out that it was something the manager had previously asked for from other departments, but up to then, had not gotten an answer.

The improvements to productivity and record keeping have also reaped benefits in some unexpected ways. For example, the city was facing a lawsuit claiming that it had neglected to fix a damaged stop sign that led to a traffic accident. Responding to this, traffic staff pulled the sign's records and showed evidence that it had actually been fixed within thirty-four minutes of when the damage was reported. Seeing the evidence, the plaintiff, who was originally seeking $250,000, dropped the case. It was great win for the city, and something that would not have been possible prior to implementing GIS for asset management.

Colorado Springs's experience demonstrates the positive impact GIS has on asset management. Today the approach sells itself within the city, and what started with signs has expanded into a citywide effort to inventory all of its assets—a true sign of success.

> The assistant city manager asked me how many 25 mph signs we had. In less than five minutes, not only could I tell him, but I could show him on a map. Andy Richter, Traffic Technician II, City of Colorado Springs, Colorado

Collaborative Utility Exchange maps Johnson County utilities

SECTOR: Utilities

Traci Robinson and Shannon Porter, Johnson County, Kansas, Automated Information Mapping System (AIMS)

Johnson County, Kansas, located in the southwest portion of Greater Kansas City, is home to a half million residents. Development expanded sharply within the county during the early 2000s, creating an increased demand for accurate utilities data. There was no shortage of data, but the methods to share and redistribute data, as well as the ability to communicate changes and plans among the various utility providers were virtually nonexistent.

Staff members working for the county's Automated Information Mapping System (AIMS) knew that a tremendous amount of time could be saved across many agencies by creating a centralized GIS system for managing utilities data. They began working toward a solution by assembling a steering committee consisting of local members of the utility community. This group was named the Collaborative Utility Exchange (CUE).

Rapid development in Johnson County created a high demand for maps and spatial data of underground utility infrastructure. Courtesy of Traci Robinson and Shannon Porter, Johnson County, Kansas, AIMS.

CUE resolved several issues concerning data sharing and standardization and paved the way for AIMS to build a GIS solution that automated the transfer of utility data from its sources to a centralized geodatabase. A secure web-based mapping portal called CUEView was then built to serve the data to the end users.

CUEView was created to provide an interactive and information-rich environment for accessing and mapping current utility infrastructure. Courtesy of Traci Robinson and Shannon Porter, Johnson County, Kansas, AIMS.

Users of CUEView have attested to its value. It saves time and money and provides a functionality that had long been missing in preliminary planning and development phases. One local water utility reported a 90 percent reduction in the time it takes to complete several common workflows—time now devoted to providing better and more efficient customer service.

One local water utility reported a 90 percent reduction in the time it takes to complete several common workflows.

CUEView puts comprehensive and current utility data directly into the hands of the people who need it and facilitates communication among the utility providers serving Johnson County. The application has helped these organizations realize a decrease in map requests coming into GIS and customer service departments while efficiency in handling client requests and overall productivity have increased.

GIS-based work order management increases productivity for the Consolidated Utility District

SECTOR: Utilities

Andy Koostra, Systems Manager, Consolidated Utility District, Rutherford County, Tennessee

The Consolidated Utility District of Rutherford County provides fresh tap water to a large portion of one of Tennessee's fastest growing counties, located southeast of Nashville. The district serves 48,500 customers and maintains 1,300 miles of waterline infrastructure.

When the district switched to a digital system for billing and work orders, Systems Manager Andy Koostra used the opportunity to incorporate GIS into the solution. It was a strategic decision that is now saving the district $3.88 per service request, which equates to $960,000 in annual labor costs.

> The Consolidated Utility District of Rutherford County, Tennessee, estimates that the increases in productivity created by marrying its electronic work order system with GIS save approximately $3.88 per service request, which equates to $960,000 in annual labor costs.

For years the district used a paper-based system to channel service requests into work orders. That process involved entering incoming requests into standardized forms that were then compiled and turned over to field crews. Each morning, the crews took their assigned stack of paperwork and headed into the field to complete their day's work.

To get crews out of the paper-based system of work orders and into a more efficient way of doing business, the district teamed with True North Geographic Technologies to develop an online GIS solution. The GIS application True North developed reads data directly from the district's enterprise geodatabase, electronic billing system, and automatic vehicle locator (AVL) to plot the location of active work orders and field crews on a detailed and interactive web map.

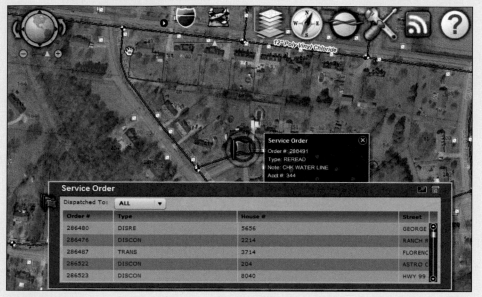

A screen shot from the GIS application created by the Consolidated Utility District to manage work orders. The GIS application ties together an enterprise geodatabase, electronic work order system, and AVL technologies to create an interactive web map that's used by field crews and customer service staff. The application has increased worker productivity by improving field logistics and removing paperwork from the process. Courtesy of Andy Koostra, Consolidated Utility District.

The district's field crews are now happily liberated from the cumbersome stack of paperwork they used to lug around. Everything they need to plan their work is now off the paper and within the online GIS application that's fed directly to tablet computers running within their vehicles. In the past it was common for work orders to get misplaced during the daily bustle, only to be found later in the day, causing the crew to double back to earlier locations. With the GIS, this problem has been eliminated—there's simply no more paperwork to lose track of. And, when new requests arise during the day, the situational awareness provided by the GIS makes it easy for the closest crew to respond while they are in the area; it is a key benefit that improves response times, which, in turn, increases the number of work orders completed each day.

Chapter 6

Generate revenue

Optimized revenue streams and cash flows sustain an organization's operational stability. In today's public sector, automated accounting, financial management, and auditing software have become staples of government finance offices. Even with these advanced approaches to monitoring the bottom line, revenue managers continued to look for more accurate methods to calculate fees and operational costs, while economic development managers began to look deeper into what types of businesses would deliver a sustainable tax base. In both of these cases—and in the areas of financial planning, grant administration, and donation collection—government agencies found that truly understanding the wealth of their jurisdiction came by tying their financial picture to their geographic outlay. For managing and generating revenue, GIS can't be beat. The technology gives financial and accounting managers an alternate way to view revenue and shows the geographic links between accounts and revenue streams.

Increasingly, financial managers are looking up from their spreadsheets and asking the question, "At what location is the revenue generated?" It's easy to see how geography enhances revenue management when you can see a link between a geographic area and successful revenue generation. Analysts can connect a specific demographic to the likelihood of service demands, to create opportunities to consolidate fee collection based on linking multiple accounts to a single location or to match capital projects to geographic demands. These demographic profiles typically cluster in specific areas. Geographies such as tax rate areas, special tax districts, sales territories, or insurance rate zones often calculate fees for service.

Traditional financial management tools and software do a good job performing routine audits and analysis, but GIS can help enhance the analysis by answering spatially related questions. The answers help identify revenues that might have slipped through the cracks. A GIS enables you to see where these "cracks" are, and enhanced revenue streams are a significant benefit of integrating GIS into traditional work flows.

Pueblo County, Colorado, grows economy with GIS

SECTOR: Economic development
Christopher Markuson, GIS Manager, Pueblo County, Colorado

According to the US Small Business Administration, small businesses have created 60 to 80 percent of net new jobs since the 1990s and employ approximately half of American workers. These facts are at the heart of the economic development philosophy of economic gardening. Pueblo County, in southern Colorado, has adopted this approach, which focuses on cultivating local businesses rather than landing large companies looking for a cheap place to do business. Instead of making a splash with thousands of new jobs coming into the community, economic gardeners favor a job here, a job there, for a slower, stable growth pattern. GIS technology is a key component in the process.

"Businesses that are already in town are not solely focused on their bottom line," said Christopher Markuson, GIS Manager, Pueblo County. "They're looking to improve business, but they're also looking to do what's right by their employees. We don't want a large company to come in, pay lousy wages, and then leave when the local economy strengthens or the workers demand higher pay."

Markuson learned about the approach from another Colorado community, the City of Littleton, when he was searching for a way to develop businesses that would not only add to the quality of life in Pueblo County but also continue to support the area during economic downturns. "We were looking at communities that rode out the last recession in the late 1990s unscathed," he said. "There were only a few, and Littleton was at the top of the list."

> To date, we've tracked over seventy-five new jobs emerging from the businesses we've helped grow, bringing over $4.6 million of new revenue into the county.
>
> Christopher Markuson, GIS Manager, Pueblo County, Colorado

The right location

Markuson took what he learned from Littleton and developed a specialized set of GIS services targeted for small businesses. Today, the Pueblo County Economic Gardening Program is a vital part of the county's GIS program and assists several businesses per week.

Businesses across the county have heard about the GIS department's consulting services, and business owners are scheduling appointments months in advance. Markuson and his small team meet with owners to find out about

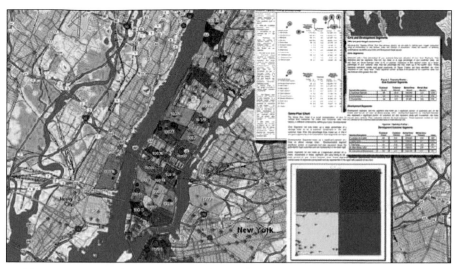

Pueblo County GIS used Esri Business Analyst to identify customer information in and around specific markets. This image of the New York City area shows sales data from a web-based business located in Pueblo, Colorado, that generates the majority of its sales from Manhattan. Markuson and his team mapped the business's market penetration by ZIP Code in darkening shades of purple, customer prospecting from yellow to red, and the amount customers spent in small-to-large blue circles. Reports from Esri Business Analyst are shown that further differentiate the business's core, developmental, and niche market groups within New York City. Courtesy of Christopher Markuson, Pueblo County, Colorado.

their concerns, interests, and current efforts. They ask questions such as, "Do you want to target advertising to reach a specific set of consumers?" and "Are you looking for a good site for a new location?" After clarifying the questions, the GIS team analyzes and maps demographic, psychographic, and other data to share with the client.

"We have this interesting reputation around town as folks that are going to give you real, truthful answers to your questions," said Markuson. "We'll tell people, 'No, the data doesn't support your plan to open a coffee shop where there are only thirty potential customers around you. It's just not feasible, and here's the evidence.'"

To get the current, accurate data for its reports and view it spatially, the GIS department uses ArcGIS for Desktop and several components and data products from Esri Business Analyst.

ArcGIS for Desktop provides a wide variety of GIS analysis and mapping, and Esri Business Analyst provides demographic, consumer expenditure, psychographic, business, and shopping center data as well as the ability to incorporate in-house data. The analysis gives business owners a thorough understanding of markets, customers, and competition. Markuson uses Esri Business Analyst Online to generate on-demand analyses, reports, and maps over the web.

"It's fast and simple," said Markuson about Esri's online GIS solutions for business analysis. "You can find employment characteristics and average daily traffic volumes for locations nationwide. We have a lot of data for Pueblo County, but not for areas outside it. With this solution, we come up with really good nationwide reports in moments—it's powerful, usable information for Pueblo businesses that are expanding into new markets."

> [Esri's online GIS solutions for business analysis is] fast and simple. You can find employment characteristics and average daily traffic volumes for locations nationwide. We have a lot of data for Pueblo County, but not for areas outside it. With this solution, we come up with really good nationwide reports in moments—it's powerful, usable information for Pueblo businesses that are expanding into new markets. Christopher Markuson, GIS Manager, Pueblo County, Colorado

A healthy harvest

The GIS team in Pueblo County works with many kinds of businesses in the area. It recently helped a local web-based business that wanted to improve market penetration nationwide. Working together, Markuson, his team, and the business owners developed strategies to increase business in fourteen of the company's top markets with advertising across media including television, radio, subway platform ads, and direct mail. They also identified the top ZIP Codes where people live who are searching for their product online and used that information to create a targeted web advertising campaign and optimize their website for search engines. The campaign is successfully bringing in new revenue for the company, and within a month of the campaign, the business created four new jobs.

"It makes sense for us to lend a helping hand to those folks that are already here, make them competitive, and basically give them the tools to expand their trade areas and reach customers more effectively," said Markuson.

Nonprofit organizations are also benefiting from the GIS department's guidance. The Pueblo Community Health Center Foundation met with Markuson and his team for less than an hour to discuss an upcoming donation campaign. Using their GIS, the team created a targeted mailing list for the campaign that resulted in a 63 percent increase in new donors.

"Christopher helped us look through different characteristics for reaching the right donors," said Janet Fieldman, chief foundation officer, Pueblo Community Health Center Foundation. By using GIS to accurately target their donors, the foundation reached its five-year fund-raising goal of $15,000 in one year. Prior to this campaign, the center purchased mailing lists based on a few demographics such as annual income and assets but had not heavily reached out to individuals because of low return and low donor acquisition. Now that it has better data and analysis, and therefore more success, the center will increase future fund-raising goals.

> **By using GIS to accurately target their donors, the foundation reached its five-year fund-raising goal of $15,000 in one year.**

"It's because they're using the right message, and there's intelligence behind who the Health Center asks for donations," said Markuson. "It wasn't who you'd think it would be—all the local philanthropists. Instead, it was the people that knew somebody who had gone to the Community Health Center for some reason."

The GIS analysis and mapping also showed the center where it should be locating services and advertising. "It allows us to decide on the right level of outreach based on the quantity of donors within particular geographic areas," said Fieldman.

A few years ago, the local community college needed to increase enrollment by 5 percent. The GIS analysis provided information it could use to most effectively market the school, which led to very positive results. The one-year goal of bringing enrollment up 5 percent increased to 17 percent, due to the targeted GIS-driven marketing Markuson and his team recommended.

"We feel we've been successful in our mission to help businesses grow and succeed," said Markuson. "To date, we've tracked over seventy-five new jobs emerging from the businesses we've helped grow, bringing over $4.6 million of new revenue into the county. I'm especially thrilled that most of these new jobs pay livable wages—$50,000 each on average, in a community whose median household wage is just over $40,000—offer benefits, and have little potential to move out of our community in pursuit of a lower-cost alternative."

Westfield, Indiana, gets fair revenues from GIS

SECTOR: Tax assessment
Eric Becker, GIS Coordinator, City of Westfield, Indiana

The City of Westfield, Indiana, is a historic place located twenty miles from downtown Indianapolis at the northern edge of the metropolitan area's suburbs. In 2008, the once-small town became an incorporated city and now has a population of over 30,000.

Shortly after incorporating, the city's administrators began to suspect that potential revenues were getting misallocated to the county, or to other cities, or simply overlooked. Staff members followed up on the issue, and by using their GIS, identified $245,000 in misallocated revenues from businesses and utilities that were owed to the city.

> By using their GIS, the City of Westfield identified $245,000 in misallocated revenues from businesses and utilities that were owed to the city.

To uncover the funds, the city targeted three potential revenue sources: taxes levied on utility companies using public right-of-way, food and beverage (F&B) taxes collected from restaurants, and franchise fees charged to customers by cable companies.

Finding the misallocated revenues and determining the proper corrections were essentially spatial problems. Staff used GIS to compare the city's corporate limits with the service areas, store locations, and customer addresses associated with these revenues, essentially creating an inventory of valid revenue sources. Staff then cross-referenced its inventories with existing tax records and boundary data to identify misallocated or overlooked revenues.

Creating a clear presentation of the analysis results was an important part of the overall process, and here again GIS played a key role. City employees used ArcGIS for Desktop to create standardized map products that highlighted the results and, as part of their presentation to the affected utilities companies, used ArcGIS for Server to publish web-mapping applications that gave companies direct and interactive access to the resultant map layers.

The maps and data generated from the GIS analysis were shared with state taxing authorities and the affected businesses. Through their work, city personnel uncovered large amounts of new revenue within each of the three areas examined. They found $70,000 in misallocated utility taxes, $65,000 in cable franchise fees, and $110,000 in restaurant F&B taxes.

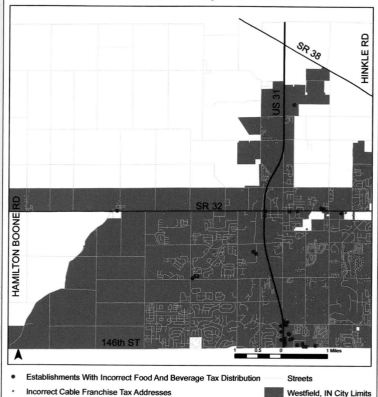

City Of Westfield, IN
Recovered Revenues From
Cable Franchise Fees
& Food and Beverage Tax Distribution

Legend:
- Establishments With Incorrect Food And Beverage Tax Distribution — Streets
- Incorrect Cable Franchise Tax Addresses — Westfield, IN City Limits

After completing their analysis, Westfield staff members used desktop GIS tools to create maps like the one shown here to clearly illustrate their findings. Courtesy of City of Westfield, Indiana.

Revenue Type	Amount
Utility Service Area Taxes	$70,000
Cable Franchise Fees	$65,000
Food and Beverage Taxes	$110,000
Total:	$245,000

As summarized in the table, the City of Westfield captured a substantial amount of new monies by using its GIS to fact check and correct the allocation of three common revenue sources. Courtesy of City of Westfield, Indiana.

GIS made what would have been a very laborious and time-consuming process efficient, accurate, and, above all, profitable. As an added bonus, the project created a baseline of data that's now used to keep tabs on revenues as the city continues to grow.

Woodstock, Georgia, adds revenue by using GIS to analyze storm water billing

SECTOR: Utilities
Emily Norton, GIS Manager, City of Woodstock, Georgia

Managing storm water runoff is an important function provided by local governments. The city of Woodstock, Georgia, has a population of 24,000 and is located thirty miles north of Atlanta. Like most municipalities, the city pays for its storm water infrastructure by charging fees to property owners. Realizing that the data used to calculate the fees was getting outdated, Woodstock implemented its GIS to update the base data for calculating storm water fees and, in the process, uncovered over $700,000 in missing revenues.

> ## Woodstock implemented its GIS to update the base data for calculating storm water fees and, in the process, uncovered over $700,000 in missing revenues.

The equation used to calculate storm water fees is based on each parcel's land use and the amount of impervious surfaces on the lot. Not having current data for this calculation meant that while the city's rapid development was putting increased pressure on the existing infrastructure, the funding needed to maintain the system and increase its capacity was staying flat.

The city decided to use an in-house and GIS-based approach to reinventory the parcel records and update land-use and impervious surfaces data. Project costs were kept low by the city leveraging its Esri Enterprise License Agreement to add more ArcGIS licenses at no additional cost and using interns from nearby Kennesaw State University as its labor pool.

The GIS department set up a data production environment centered on a multiuser geodatabase. The interns overlaid the parcels with high-resolution imagery, then updated the land use and impervious surfaces data for each parcel in the city. When necessary, teams conducted field checks to verify their updates and to collect land-use and land-cover data for new developments not shown in the imagery.

Using a team of interns, a multiuser geodatabase, and ArcGIS maps like the one shown here, Woodstock's GIS program set up a production environment for updating the map layers used for calculating storm water fees. It was a low-cost project that produced big returns by uncovering hundreds of thousands of dollars in additional storm water revenues. Courtesy of Emily Norton, GIS Manager, City of Woodstock, Georgia.

Revenues Located	Amount
New Storm Water Fees	$206,000
Re-Allocated Property Taxes	$495,000
Total Revenues Gained:	$701,000

Woodstock's GIS project led to substantial increases in revenues, providing another solid example of how using GIS to solve problems and aid analysis can go a long way toward keeping a city above water in its efforts to keep pace with development. Courtesy of Emily Norton, GIS Manager, City of Woodstock, Georgia.

The entire project was completed in fewer than three months at a total cost of just under $25,000. Once completed, the updated data was plugged back into the billing equation, which showed an increase of $206,000 in the total revenue collected from storm water fees. In addition to the increase in storm water revenues, the GIS project also located 345 parcels within the city whose property taxes were incorrectly allocated to the county. It was a substantial finding that netted the city an additional $495,000 in revenue.

Chapter 7

Increase efficiency

GIS is an invaluable tool for increasing productivity without waste in business operations. When government business units increase efficiency, they realize improvements in the delivery, enhancement, and more effective use of a service. Many organizations periodically rethink and reengineer their business workflows to improve the effectiveness of processes and advance their overall mission. These efforts usually result in improved service delivery by eliminating redundant or outdated steps in old processes, finding ways to alter or reduce staff workloads and trips, and developing new and innovative procedures. GIS reengineering or georeengineering requires management to look at the role geography plays in its business processes or how geography can act as an interface to deliver a service.

A common GIS reengineering or georeengineering application involves the movement of employees in the field, which generally means routing workers within specified time constraints. Another frequent business problem is how to give employees access to a knowledge base from which they can retrieve information in the course of answering customer inquiries or making other business decisions. GIS can help resolve or alleviate these kinds of dilemmas by linking all or most business data with its location. Transferring these concepts into GIS applications of Government 2.0 or web-based services can act as an extension of staff delivering customer service twenty-four hours a day, seven days a week.

When workers collect data in the field that is related to a specific location, the implementation of a mobile GIS will enable them to travel with geographic-related data and reduce or eliminate repeat trips back to the office for information or files. Data collected in the field is accurately entered once and uploaded into the main database for immediate use.

A good approach to considering ways to increase efficiency with GIS is to list all tasks that require reengineering and then identify their geographic components. This strategy leads to increased savings of staff time and resources while achieving greater productivity and speed.

Golden, Colorado, does more with less by using GIS for sign management

SECTOR: Public works asset management
Quintin Pertzsch, GIS Coordinator, City of Golden, Colorado

The city of Golden, Colorado, lies ten miles west of Denver in the foothills of the Rocky Mountains. This scenic city is home to 18,000 residents and a Public Works Department that gets big returns by using GIS to help maintain its street signs.

Like most communities, Golden is continually looking for better and more efficient ways to do its work, and it has been using Esri's GIS technologies to do so since 1999. Initially, the Public Works Department used GIS to help map its utility infrastructure, but over time the department began to use GIS within other applications, such as the management of the city's road signs.

Prior to using GIS, Golden's street signs were managed on an ad hoc basis. There was no regular inspection or replacement schedule for signs, and they were repaired on an as-needed basis. It was a reactive program that made it nearly impossible to adequately account for maintenance costs or even to know how many signs they were responsible for. Golden found the solution to its sign management woes in GIS, and the solution was so successful that GIS Coordinator Quintin Pertzsch described it as "nothing short of amazing."

Inventory creates awareness

Golden's first step toward a GIS-based approach for sign management was the creation of an accurate inventory. A data-collection project was implemented utilizing GPS with data loggers to record sign locations and their primary attributes. Crews visited each sign in the field, marked their location with the GPS, and logged each sign's attributes. These attributes included the MUTCD (Manual on Uniform

> Golden found the solution to its sign management woes in GIS, and the solution was so successful that GIS Coordinator Quintin Pertzsch described it as "nothing short of amazing."

Traffic Control Devices) Code, sign material, support characteristics, and facing direction. Crews also recorded the condition of each sign during the data-collection work.

This field work resulted in an accurate and detailed digital inventory of the total stock of signs in Golden. The information was stored as a map layer within the GIS, making it possible for staff to quickly render various types of sign maps. City workers used the maps to answer questions about sign inventory and to help plan their field work. They could map all the signs in the city, only signs of certain type, or just those signs that met a specific set of criteria, like speed limit signs in poor condition; and because the data is all stored as tabular information in the GIS, they have the ability to generate reports that summarized the information from various queries.

A City of Golden employee inspecting signs with the mobile GPS unit. Courtesy of City of Golden, Colorado.

The city used its new inventory to create an inspection schedule based on the sign classes. Regulatory signs, such as stop, yield, and speed limit signs, were scheduled for yearly inspections, and nonregulatory signs for once every three years. Even though a tremendous amount of good information about the signs had been recorded and stored within the GIS, inspections and maintenance were still driven by a paper-based approach that was time-consuming, inefficient, and confusing—not to mention difficult to read. Furthermore, the field crew would turn in its completed paper-based inspections and data-collection sheets to data-entry employees who had to decipher what was on the sheets. It became clear that, while the GIS-based sign inventory had tremendous value, much more work was needed to improve workflows.

Return on investment

The new sign inventory led to several initial benefits, but the ultimate goal was to leverage the GIS data to improve maintenance practices while also reducing operating costs. Specifically, the city wanted to improve the timeliness of inspections, eliminate unnecessary paperwork, and

A map showing failed street signs in Golden that need to be replaced based on recent inspection. By viewing this information in the GIS, an accurate route can be determined, as can the correct number and type of sign that need to be brought into the field. Courtesy of City of Golden, Colorado.

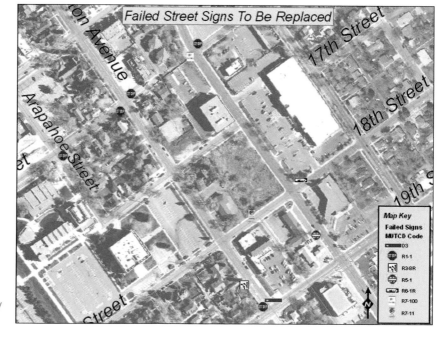

Failed Street Signs To Be Replaced

Map Key
Failed Signs
MUTCD Code
D3
R1-1
R3-8R
R5-1
R6-1R
R7-100
R7-11

> Before using GIS to assist in the maintenance program, the average sign inspection took fifteen minutes and cost $5.27. One year after implementing GIS, the average inspection took only three minutes and cost $1.54.

provided the additional functionality the city needed to fully integrate its burgeoning sign maintenance program with GIS.

After setting up the system and developing a mobile-based approach for sign maintenance, the city began to track its inspections and maintenance work for the active stock of more than 5,000 street signs. In so doing, the city not only tracked what signs were inspected, but also the time and the cost of each inspection.

Using the information coming in from the new system, the city was able to create an overall baseline of annual inspection costs that was useful for comparisons. Before using GIS to assist in the maintenance program, the average sign inspection took fifteen minutes and cost $5.27. One year after implementing GIS, the average inspection took only three minutes and cost $1.54. The city had achieved its goals to improve efficiencies and lower costs.

The baseline comparisons helped Golden clarify its total return on investment from GIS, especially when examined on an annual basis. For example, in the year before using GIS to inspect signs, the city estimates that it cost $11,400 to inspect 2,177 signs. One year later,

increase the efficiency of workflows. To help reach these goals, Golden implemented Cartegraph's asset management tools for street sign maintenance. Cartegraph's tools worked directly with Golden's existing GIS and

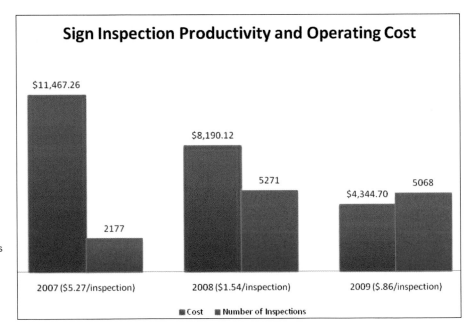

Sign Inspection Productivity and Operating Cost

As the graph shows, the efficiencies gained by implementing mobile GIS technology for sign inspections have sharply decreased costs and increased productivity.
Courtesy of City of Golden, Colorado.

$11,467.26 — 2007 ($5.27/inspection) — 2177
$8,190.12 — 2008 ($1.54/inspection) — 5271
$4,344.70 — 2009 ($.86/inspection) — 5068

■ Cost ■ Number of Inspections

with the use of GIS, the cost to inspect 5,271 signs was only $8,000. The mobile GIS solution made it possible for Golden to inspect more signs for less money.

The substantial savings over the first year led managers and staff to wonder if the results were a fluke or something that could be repeated. To their pleasant surprise, the results were even better the second year, as the costs per sign inspection had dropped to a mere 86 cents.

The cascading effect of efficiency

Golden's use of GIS for sign management has created many benefits. Today the Public Works Department systematically inspects each sign in its inventory at least once a year, which means that it knows at any given time the overall condition of its stock and can efficiently and proactively carry out repair and replacement work. It has also received a logistical boost in its field work, which has improved its responsiveness to work orders while reducing fuel costs through better route planning. The system has even reduced legal liabilities for the city because each inspection is fully documented and stored with its associated sign record in the GIS.

Perhaps one of the biggest benefits of this use of GIS is how much easier it is for Golden to conform to the new sign regulations set by the Federal Highway Administration (FHWA). The FHWA has defined a rigid set of compliance specifications for signs that all local governments must conform to incrementally over a series of deadlines. The accurate sign inventory now housed in Golden's GIS is making it easy for the public works staff to pinpoint signs that do not meet the new standards and strategize their work to reach full compliance with the FHWA standards on time and within budget.

Golden is also drawing from the intelligence provided by the GIS-based sign inventory to help weed out unnecessary or redundant road signs. Doing this improves the attractiveness of roadway corridors by reducing visual clutter, and importantly, each sign removed is one less sign to maintain and pay for. It is something that residents and visitors benefit from even though they probably do not even know that it has happened, and it is yet another example of something Golden could not have easily accomplished without the big picture and analytical capabilities provided by GIS.

Irish councils automate data collection with GIS to improve public feedback processes

SECTOR: Regional planning
Colette Cronin, GIS Officer, Dún Laoghaire-Rathdown County Council, Ireland
Hazel Farley, Systems Analyst, GIS Team, Fingal County Council, Ireland

Dún Laoghaire-Rathdown (DLR) County Council and Fingal County Council are both based in County Dublin, Ireland. In the wake of the global economic crisis, councils are having to tighten their budgets and work more efficiently while still improving services for citizens. These two enterprising local authorities have done precisely that by cooperating on the development of a new online GIS application used to gather information about county development plans.

Across Ireland, every county and city council is obliged to produce a county/city development plan every six years to set out an overall strategy for development in its region. During the production of these development plans, councils have to provide three formal opportunities for public consultation. The two councils wanted to make it easier for the public and other interested parties to participate in this consultation process by enabling them to view and submit information online in connection with development proposals.

GIS teams and development plan personnel from both councils worked closely together with consultants from Esri Ireland to design, develop, and deploy an online solution based on ArcGIS. Today, the solution is used by members of the public to enter a summary of their concerns, make a full text submission, mark the boundary of the area relating to their submission directly onto the plan, and attach up to five supporting documents. When a submission is completed, the solution automatically enters the data into the council's back office systems. "It removes a substantial amount of work that our development plan team would otherwise have to undertake internally," said Hazel Farley, Systems Analyst at Fingal County Council.

The Fingal County Council estimates that the GIS solution saves ten to fifteen minutes for every web submission received. Considering that the first phase of public consultations could generate up to 1,200 submissions, the time saved by automating data collection with online GIS can potentially save sixty to eighty hours per round.

The GIS solution saves ten to fifteen minutes for every web submission received. Considering that the first phase of public consultations could generate up to 1,200 submissions, the time saved by automating data collection with online GIS can potentially save sixty to eighty hours per round.

Speaking about the solution, Colette Cronin, GIS Officer at the DLR, said, "The council achieves savings from reduced workload during a very busy time period, due to reduced queries from the public and more efficient working practices. The public, in turn, experiences reduced costs by being able to view and interrogate the maps at home, rather than having to travel to our offices or the local library to look at them."

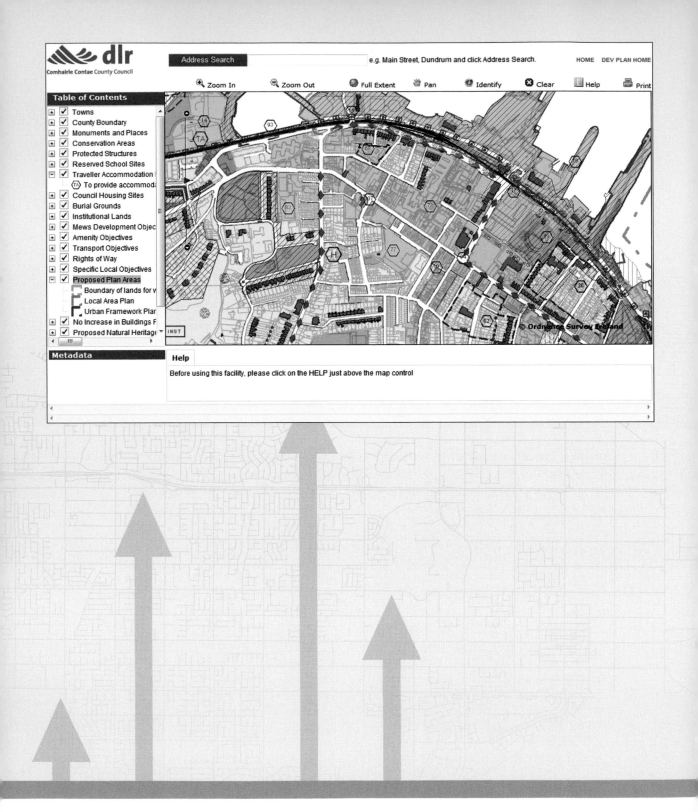

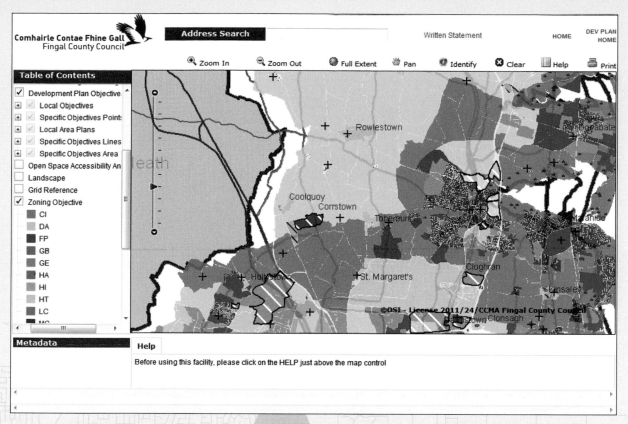

The DLR and Fingal County Councils collaborated to create the online GIS applications shown here to help automate and streamline the mandatory public review process of development plans. The applications are popular with members of the public, who can submit their feedback directly from the comfort of their homes. They are also popular with both councils, because they have streamlined the data-collection process, which is saving valuable time.

Map data courtesy of Ordnance Survey Ireland. ©Ordnance Survey Ireland/Government of Ireland. Copyright Permit No. MP 005911.

Mobile GIS improves code enforcement services in McAllen, Texas

SECTOR: Code enforcement

Jose J. Peña, Application Services Manager, City of McAllen, Texas

Municipal code enforcement is no easy job, but it is a critical function that helps keep neighborhoods clean and livable. The City of McAllen, along the Rio Grande in Texas, takes code enforcement seriously, and when the budget for the work tightened as the workload increased, this growing city of 130,000 residents turned to a mobile GIS solution.

For years, code enforcement staff had used paper-based processes to issue orders and plan and carry out work. Each day before going into the field, officers spent valuable time in the office checking several sources to get all of the background information on the cases organized and their paperwork in order.

After examining their workflows for ways to improve them, the staff automated the code enforcement processes with a mobile GIS solution that tied Accela Mobile Office to real-time situational-awareness maps that were distributed from a centralized ArcGIS for Server.

Code enforcement officers now use GIS in the field to display, inspect, capture, and update geographic information related to their cases. The solution has connected code enforcement databases to parcel data and permitted real-time information sharing between field crews and office staff. Cases are now digitally tracked from start to finish, which drastically increases the completion rate of cases because nothing can get lost in a mountain of paperwork anymore.

Speaking about the new GIS-based way of doing business, Project Manager Jose J. Peña said, "Now the code enforcement process is more accurate and efficient because everything is available in one solution rather than having to go to different sources for the data."

Prior to using GIS, officers were completing roughly 5,000 code enforcement cases annually. Today, with the aid of GIS, the city is conducting approximately 23,000 inspections a year to the tune of 16,000 completed cases annually.

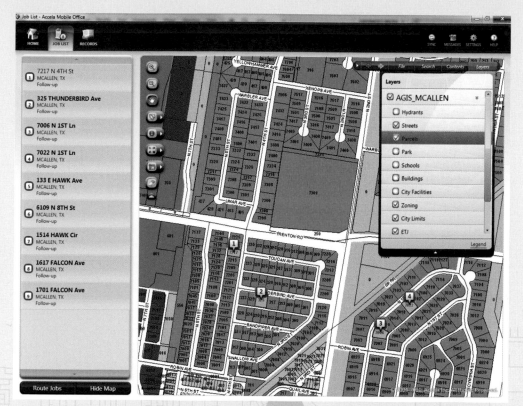

McAllen's mobile GIS solution for code inspection organizes several information sources into a single map-based framework for carrying out and documenting work. Code officers use the GIS in the office and the field to gather information, document inspections, map cases, and route trips. The GIS solution has wiped away stacks of paperwork and dramatically increased workflow efficiencies, giving McAllen a 300 percent increase in the number of code violation cases successfully resolved each year. Courtesy of Jose J. Peña, Application Services Manager, City of McAllen, Texas.

The efficiency gains from McAllen's GIS solution for code enforcement have improved response times and dramatically increased the number of cases officers complete each year. For example, prior to using GIS, officers were completing roughly 5,000 code enforcement cases annually. Today, with the aid of GIS, the city is conducting approximately 23,000 inspections a year to the tune of 16,000 completed cases annually. More completed cases means cleaner neighborhoods and a more livable city.

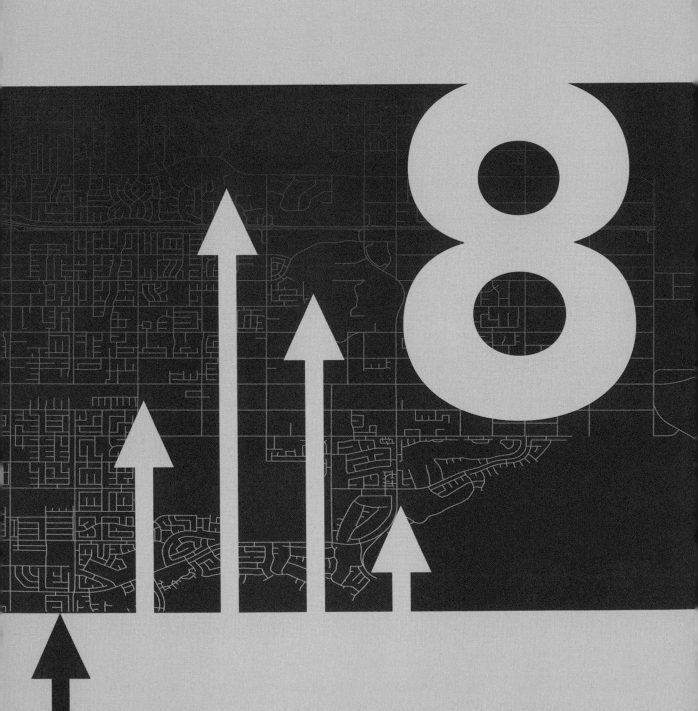

Chapter 8

Automate workflows

From a management perspective, GIS generates widespread benefits efficiently and productively. Linked to business processes, GIS offers an organization the opportunity to automate a variety of tasks that expedite workflows and enhance an organization's ability to react efficiently during a crisis.

Managers can find opportunities to automate processes with GIS via system integration in enterprise resource planning, mobile workforce automation, web-centric mobile applications, and customer-relationship management systems. Managers can use GIS to automate routine analysis, map production, data creation and maintenance, reporting, and statistical analysis. Often, geoprocessing can yield significant reductions in task replication. Mobile GIS can reduce the number of trips to the office for data retrieval, and it offers a data-collection avenue to eliminate steps between the field and traditional key entry. Internet-based solutions can cross-reference data repositories and deliver a self-help approach to perform many tasks that traditionally require direct personnel involvement. GIS can automate repetitive tasks seamlessly so that nontechnical staff can maximize productivity while not feeling intimidated by the software. Many GIS applications are readily available to organizations out of the box or as solutions from third-party developers.

GIS operates on a reusable data foundation that interacts and supports numerous business systems and information bases. This data reuse and reassignment of the information to different problems is unique to GIS technology because GIS can communicate with back- and front-office systems and access real-time data and information. Grounded in geography, GIS offers an exceptional method for performing business processes, unlike traditional data-organization methodologies.

The Virginia Department of Forestry uses GIS to automate workflows in office and the field

SECTOR: Forestry
Timmons Group

The Virginia Department of Forestry (VDOF) manages more than fifteen million acres of land. To help protect and sustain its forestland, VDOF is continually collecting and mapping data about the condition of Virginia's state forests. VDOF is a large organization with an annual operational budget of $26 million and over 250 salaried staff. Considering its size and the scale of its work, it comes as no surprise VDOF has turned to GIS to assist its mission. Today, through the use of GIS solutions, VDOF is connecting workflows, automating tasks, and saving thousands of dollars annually by eliminating redundant costs.

For years, VDOF used manual methods to track and communicate information about its work. Crews went into the field with paper maps and notebooks, logged forest conditions and management information by hand, and then brought the paperwork back to the office where it was reentered into tabular databases. After the initial work was completed, the hard-copy maps were then stored in file cabinets at county offices around the state.

The manual nature of work at VDOF had several limitations and inefficiencies. Having maps and databases scattered across offices throughout the state made the task of compiling a complete picture of forestry activities within Virginia very difficult and time-consuming. And just as important, the ability to compile the history of management practices for any given area—which is critical to decision makers within VDOF—was hindered by the old manual process.

VDOF also recognized the need to modernize field work by creating more efficient data-collection methods, including a direct connection between the field and office. Essentially, VDOF wanted to liberate field crews from lugging around hard-copy maps and other paperwork, automate field data collection, and provide a way to perform many standard business workflows directly in the field.

Recognizing the need to improve workflows, VDOF partnered with Timmons Group, a civil engineering company with a large technology and geospatial services practice, to develop a web-based GIS to modernize its processes. Working together, they created a GIS solution called the Integrated Forest Resource Information System (IFRIS).

During the first phase of the IFRIS implementation, the development team created an enterprise GIS that centralized its cache of spatial data within a geodatabase. The team then constructed a set of web-based data automation tools used to edit, validate, store, and manage this data. A key part of the solution is functionality that automatically tracks and archives changes made to the spatial data within the geodatabase, essentially letting a user trace the history of forestry data for a given location.

The enterprise solution was developed with ArcGIS for Server and uses Microsoft SQL Server for its database platform. The solution uses web-GIS technology to feed data from the centralized geodatabase to approximately one hundred remote offices distributed across the state.

The Virginia Department of Forestry (VDOF) uses a centralized and web-based GIS solution known as the Integrated Forest Resource Information System (IFRIS) to assist in its forest-management projects. IFRIS is a comprehensive application that has improved data management, automated several key workflows, and created a direct connection between field operations and office work. Courtesy of Timmons Group.

Forestry staff at the remote offices access IFRIS through the web and use the application for a variety of purposes, such as defining the location and boundaries of forestry activities, updating information about forest health and pests, reviewing data already in the system, and simply creating maps.

After the initial phase of IFRIS was completed, the application was expanded to include mobile functionality that created a direct link between the field and the office along with the ability to conduct standard business workflows directly from the field. The tools were designed to standardize collection, reduce time, eliminate paper work, and lower the expense of fieldwork.

The Timmons Group, working on behalf of the VDOF, and using Esri's mobile GIS technologies, designed the solution to run on Trimble GPS devices with a standard set of forms used for data collection and entry. Automation routines were also created to track and report staff time and staff accomplishments directly from the field.

Automation was at the center of the mobile solution. Information collected in the field was automatically tied to its real-world coordinates. Digital forms were designed to speed data entry and standardize the information collected. To transfer the data back to the office, web services were programmed to automatically push data from the mobile GPS devices back to the enterprise geodatabase, making it available to all IFRIS users nearly in real time.

Through increases in data access and data quality, the use of GIS technologies has improved VDOF's overall awareness about ongoing forest-management programs.

The VDOF uses its web-based GIS application for a variety of purposes such as defining the boundaries of timber stands, updating information about forest health, and simply reviewing data and printing maps. By establishing a centralized GIS with tools and data that are shared via web services, the application and its data are easily distributed to offices across the state. Courtesy of Timmons Group.

IFRIS leverages mobile GIS and GPS technologies to automate the field collection of forestry data and track the work of field crews. The mobile application has streamlined workflows, eliminated several paper-based forms, and reduced operational costs. From Shutterstock, courtesy of Melissa King.

Datasets that used to be scattered across the state at county offices are now accessible to VDOF staff through a single web portal. Regardless of their location, staff members can access the data they need, as they need it. The enterprise GIS also eliminated a long-standing time lag between data entry and availability for use within a wider forest-management context.

The mobile GIS capabilities within IFRIS put geospatial tools directly into the hands of field crews. It increased the speed and ease with which workers capture, map, and describe forestry information. The integration of GIS and GPS for field data collection has automated the direct transfer of information from the field to the office, where it's immediately available for other uses.

The workflow automation provided by IFRIS also makes it possible for crews to monitor field work and fill out appropriate time sheets without having to return to the office. The system has reduced the amount of paperwork and automated several data entry processes.

The GIS has cleared away the limitations of the previous paper-based system through the automation of common routines, and today forestry data at VDOF is ready to use right after it is collected in the field. This gives regional managers the ability to monitor the details of department projects and make rapid adjustments that target service resources to high-priority areas.

As much as IFRIS benefits the day-to-day work of forest management, the enterprise GIS was also a sound financial decision. After comparing the costs of IFRIS to a

> **VDOF estimated that going with a centralized GIS saved more than $80,000 in initial software and hardware costs, and just over $45,000 annually in licensing and maintenance costs.**

decentralized desktop solution that had had been previously considered, VDOF estimated that going with a centralized GIS saved more than $80,000 in initial software and hardware costs, and just over $45,000 annually in licensing and maintenance costs.

Today, IFRIS is fully operational and its uses have expanded to more specialized areas such as fire tracking and water-quality analysis. The widespread use of its enterprise GIS is leading to better mapping of forest-management activities and much richer datasets than the agency and its customers had prior to the implementation. Because of these benefits, field staff members are making better and more informed decisions about how landowners should manage their resources—decisions that pull from the historical intelligence provided by the IFRIS geodatabase—to consider both short- and long-term goals.

> **The GIS has cleared away the limitations of the previous paper-based system through the automation of common routines, and today forestry data at VDOF is ready to use right after it is collected in the field.**

EastLink Tollway: GIS puts road construction project in the fast lane

SECTOR: Construction

Jason Clark, GIS Manager, Thiess Pty. Ltd.

Construction of the EastLink Tollway in Melbourne has been billed as one of Australia's largest road projects. The project spanned forty-five kilometers, included seventeen interchanges, ninety bridges, and a twin tunnel dug through 1.6 kilometers of earth.

Four months after the first stone was turned, project staff identified GIS as an ideal solution to host the myriad of location-based and textual data produced by a project of EastLink's size. The staff rapidly developed a GIS to automate mapping and information flows that helped reduce the time to complete the project and save thousands in costs.

> The staff rapidly developed a GIS to automate mapping and information flows that helped reduce the time to complete the project and save thousands in costs.

"We knew that our GIS solution needed to provide high-quality cartographic capabilities, supply timely and accurate data to stakeholders, integrate with other corporate systems, be simple to use, be put into service by existing staff, and—most important—be cost-effective," said Jason Clark, GIS Manager on the project. Clark also noted that the solution had to be interoperable with the computer-aided design (CAD) files produced from within the design and engineering processes.

The project staff teamed with Esri Australia to create an innovative, multistaged GIS strategy. The implementation included the procurement of new office and field hardware and software, the translation and integration of CAD data into a geodatabase, and the development of web-enabled applications for publishing data across the project's enterprise.

Mobile GIS solutions were used to strategize field data collection and monitoring. Customized forms running on mobile GIS devices were created for tunnel face mapping, and ArcGIS 3D Analyst was used to create water-monitoring models throughout tunnel construction.

The GIS solution delivered numerous benefits that included a 50 percent improvement in map production, automated spatial data workflows, faster access to timely and multilayered data, and improved reporting capabilities. The automated data validation and quicker editing of features helped reduce erroneous data entry, ensuring more accurate data, while the time required to locate and collate information in the system was dramatically reduced by 80 percent. Overall, the GIS strategy assisted in cutting EastLink's construction time from forty-eight to forty-two months and contributed to significant savings in project costs.

GIS Benefits EastLink Tollway Project
Multiple Solutions GIS Services provide multiple solutions despite limited resources.
Automated Worklows An automated workflow was managed in a centralized and consistent format.
Better Maps, Faster Cartographic map production was 50 percent faster.
Money Saved Project cost avoidance and savings exceeded $100,000 per year.
Increased Situational Awareness Faster access to accurate and timely data was achieved.

As shown here, GIS provided numerous benefits and a significant return on investment to the EastLink Tollway Project. Courtesy of Esri Australia.

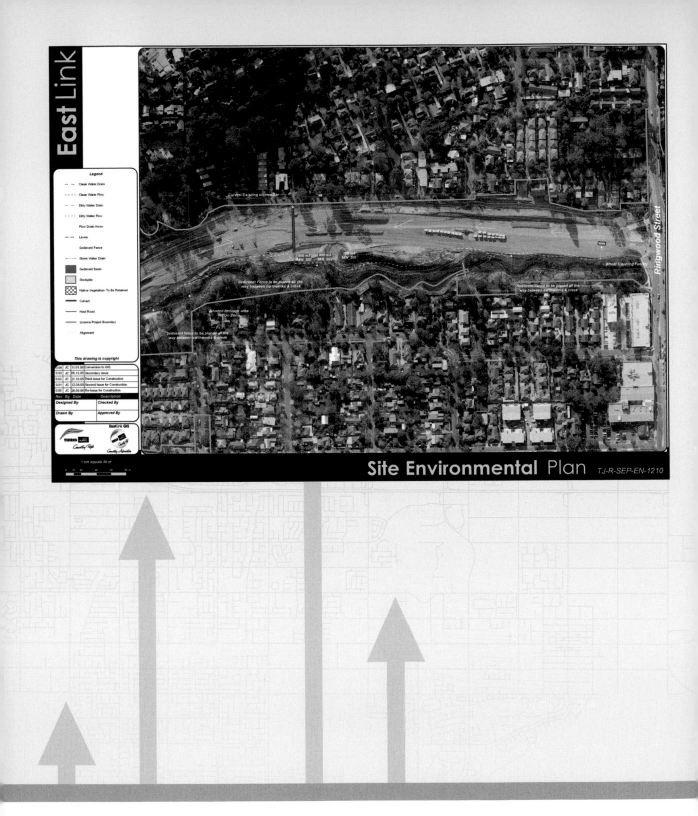

Site Environmental Plan   TJ-R-SEP-EN-1210

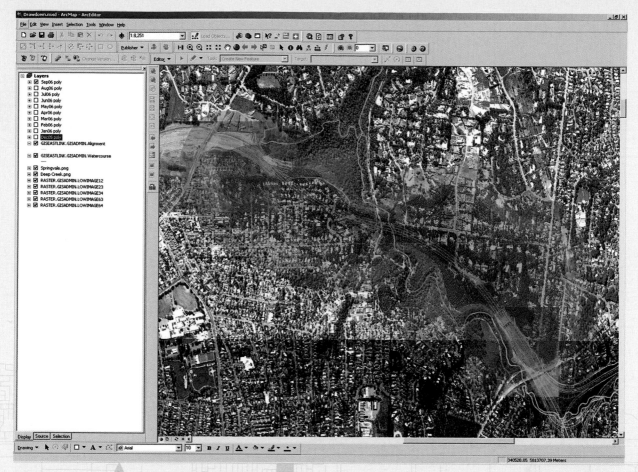

GIS was an information hub for the EastLink project. It was used for a variety of purposes, from automating mapping and data entry to groundwater modeling during tunnel construction. Courtesy of Jason Clark, Thiess Pty. Ltd.

Nova Scotia Power saves $200,000 in labor costs through GIS field connectivity

SECTOR: Electric

Brian Shannon, GIS Connectivity Project Manager, Nova Scotia Power Inc.

Electricity distribution networks carry power from the transmission system to the consumer—be it industrial, commercial, or residential. To provide reliable customer service and minimize voltage quality issues, a utility must develop an accurate model of its electric distribution system to ensure that it has precise location information about customer connectivity and utility assets.

With nearly half a million customers to service, Nova Scotia Power Inc. (NSPI) embarked on a mission to improve its electric distribution system model through mobile data collection. Looking to enhance productivity, save time, and improve data accuracy, NSPI initiated a three-year GIS Connectivity Project aimed at providing connectivity in the field for immediate GIS server updates. The project solution allowed the utility to build a reliable electric distribution model to better predict the impact of storms, to dispatch line crews more effectively, and to provide precise information to customers and emergency officials during outages.

NSPI used mobile GIS to collect field data about its utilities infrastructure. The asset information captured in the field was automatically uploaded and integrated into NSPI's GIS applications. From Shutterstock, courtesy of Payless Images.

With more than 34,000 square miles to cover, one of the utility's biggest challenges was completing the project on time. Data collection was to occur by foot and vehicle in urban rural and off-road areas, and needed to be completed within a three-year time frame. To make it happen, the utility equipped data collectors with BlackBerry smartphones and Freeance Mobile software that enabled the phones to connect directly with NSPI's network.

By eliminating the need for data collectors to travel to the local utility depots to upload data, we saved two hours a day and nearly Can$200,000 in labor costs. Brian Shannon, GIS Connectivity Project Manager, NSPI.

By using a mobile solution for data collection, field data collectors could capture and publish information directly from a BlackBerry without having to return to the office to upload and share information. Data was automatically transferred to Esri's ArcGIS for Server technology, so that technicians could update the electrical model within minutes.

"By eliminating the need for data collectors to travel to the local utility depots to upload data, we saved two hours a day and nearly Can$200,000 in labor costs," said Brian Shannon, GIS Connectivity Project Manager, NSPI. "We've built an accurate picture of our distribution system to better manage our Can$3.5 billion in utility assets and provide customers across the province with reliable service."

Chapter 9

Manage resources

Efficient resource management means analyzing, tracking, managing, allocating, and conserving assets. Enhanced resource management might involve influencing the way a government uses its assets, or it can take on a broader context to include managing the earth's precious commodities to sustain our world. Applications of resource planning occur in all disciplines.

For quick and efficient production and delivery, many governments have cultivated enterprise resource planning technology. This technology helps to streamline, analyze, and automate business processes such as accounting, inventory, e-commerce, human resources, sales, shipping, and customer service. These companies are integrating GIS with their resource planning technology systems to maximize benefits and effectively manage resources.

In an emergency, utilities must quickly mobilize staff, equipment, and supplies. Integrating various systems and work processes streamlines the effort. GIS leverages geography to generate a common operating picture that clearly defines where resources are located and where they need to be allocated. The customer-relationship management system helps to collect customer calls related to an outage; the outage management system identifies the point(s) of failure; enterprise resource planning ensures the dispatch of necessary resources; and the GIS integrates information to optimize travel routes.

GIS, integrated with enterprise resource planning and business processes, benefits businesses and governments as a decision-support tool that sustains our earth and impacts costs, productivity gains, inventory control, and staffing. In essence, widespread GIS adoption has led to the rise of map-centric interfaces that are now the executive dashboard of this generation of management.

The success of resource management depends on understanding how geography and its related elements affect commodities. Start by asking where assets are located. Then look for the origin of the demand for these assets, or which local issues or climates affect the resource.

Government workers have connected data collection, analysis, and logistics to the executive dashboard in scenarios ranging from hurricane response to tragic oil spills along coastlines. The public health industry is using GIS to relate location-based demographics, such as the location of at-risk populations, to the allocation of home health care providers. Agribusiness uses GIS to study crop yield by location to understand the logistics of getting products to market.

Masdar City relies on GIS to help create one of the world's most sustainable urban developments

SECTOR: Sustainable development
Derek Gliddon, GIS Manager, Masdar, Abu Dhabi, United Arab Emirates

In the heart of the Arabian desert, hundreds of dedicated people, billions of dollars, and years of effort are coming together to achieve a groundbreaking goal: Masdar City, one of the world's most sustainable urban developments. It is an immense project in which GIS is providing critical intelligence to manage resources and target goals.

The vision of a sustainable city

Masdar City, in Abu Dhabi, United Arab Emirates, aims to be to the clean technology sector what Silicon Valley is to the IT sector. Commissioned by Masdar, Abu Dhabi's renewable energy company, the city's six-square kilometers will eventually be home to 50,000 people and some 40,000 daily commuters, and will use renewable energy sources and energy efficient architecture. To aid its goals, Masdar City is using GIS to help lower the use of critical resources throughout the development lifecycle.

The construction of Masdar City is a project of colossal proportions, with no historical frame of reference. Yet, it is also a commercial venture with savvy project managers who have turned to Esri's GIS technologies to assist in its design and building, manage costs, and provide long-term monitoring tools that make it clear to the rest of the world that Masdar City is achieving its goal as one of the most sustainable urban developments in the world.

Geodesign for a low-carbon city

On the Arabian Peninsula, understanding the opportunities and constraints of the local geography, regional cultural norms, climate, and natural resources is vital for success. Masdar City is being planned with careful consideration of human and physical geography, economics, and social perspectives. Sun angles, wind patterns, street widths, building density and height, and even city

GIS is being used to help reduce the use of critical resources from city design to operation.

A conceptual aerial view of how Masdar City will look when completed. *Courtesy of Derek Gliddon, GIS Manager, Masdar.*

GIS has proven invaluable by providing planning and logistical support for everything from generating optimal utility routes to determining the placement of construction materials to locating facilities for large numbers of workers. In each of these areas, project staff uses GIS to compare alternatives and pick the most efficient solutions to save money and reduce carbon output.

orientation are all being studied using GIS. A substantial task associated with the design of Masdar City is optimizing the placement of a wide variety of renewable energy and green-technology facilities such as water and sewage treatment plants, recycling centers, a photovoltaic solar array, geothermal wells, and productive landscapes used to create locally grown crops. The approach taken by the design teams combines traditional planning principles with GIS to optimize the placement of city elements within the site. It is essentially an exercise in geodesign that is used to solve spatial questions such as: Is there enough physical space? How do we maximize shading of buildings and encourage natural breezes—yet ensure personal privacy? How much space do we need between a facility and residents? To answer each of these questions, Masdar's design teams use GIS to select sites, evaluate alternatives, and visualize the construction sequences.

GIS at Masdar City will be completely integrated into daily operations and fully leveraged from day one. It is essentially woven deep into the fabric of technology that is in place to manage and sustain the city's natural and man-made resources.

Masdar design teams use GIS to model building information throughout the life cycle of the project, which will then be used post construction to help manage the city's infrastructure and sustainability goals. Courtesy of Derek Gliddon, GIS Manager, Masdar.

GIS played a critical role in the design and build of Masdar City. Courtesy of Derek Gliddon, GIS Manager, Masdar.

Speaking of construction, to be a truly low-carbon city, it, too, must be engineered to ensure minimal waste and optimal use of resources. This is a considerable challenge for a project this size that's located in desert terrain, and here, too, GIS has proven invaluable by providing planning and logistical support for everything from generating optimal utility routes to determining the placement of construction materials to locating facilities for large numbers of workers. In each of these areas, project staff uses GIS to compare alternatives and pick the most efficient solutions to save money and reduce carbon output.

While creating a sustainable city is the main aim of this project, it must be effectively balanced with cost. Managing the cost of building Masdar City on a daily, weekly, and monthly basis is extremely important to sustaining the project's momentum. The Masdar GIS team built a GIS-facilitated 3D visualization of the construction costs, carbon emissions, and schedule to help program managers understand the status of these continuously changing performance measures.

Modeling a city and its assets with GIS

Masdar City is in a uniquely beneficial situation because it and its GIS are being built in tandem. Because of this, GIS at Masdar City will be completely integrated into daily operations and fully leveraged from day one. At the heart of the development's information infrastructure, GIS is a key part of the integrated technology suite that is in place to continuously improve the city's sustainability.

An example of the integrated aspect of GIS and spatial data at Masdar City is how project design teams are using building information modeling (BIM) to provide 3D models of all structural components of the city. They are essentially creating a digital model of the city's built environment that plugs directly into the city's GIS.

To complement BIM, export routines are being built to feed Masdar's model data into the city's geodatabase to locate every single asset with pinpoint precision. This includes the location and interrelation of all electrical and ICT (information and communications technology) cables, water and sewer networks, and the transportation infrastructure. When the city is fully operational, this will enable simpler, more effective asset management. This means that Masdar City will be able to do proper incremental maintenance on its assets to ensure that all systems are running at peak performance at all times.

Masdar's GIS will also be integrated with a computerized maintenance management system (CMMS), which will automatically generate work orders that are sent directly to technical engineers who will then carry out the work and instantly update the system. The whole process will be paperless, contributing further to a sustainable city.

By integrating with Building Management Systems, GIS will help city management analyze the resource use and carbon balance within every room of every building. Masdar will use GIS to visualize energy and water usage for the city as a whole and communicate this information back to residents in new and novel ways.

> By raising awareness, building in incentives, and communicating the real-time status of the city's sustainability, GIS will play a major role in helping everyone in Masdar City work together to cut down on resource use.

> Design teams working on Masdar City have stated that the project work and project goals would be infinitely more difficult without GIS and its data management, integration, sharing, analysis, and visualization capabilities.

Each resident will be able to see a more detailed breakdown of information—such as how much water is used during a shower. By raising awareness, building in incentives, and communicating the real-time status of the city's sustainability, GIS will play a major role in helping everyone in Masdar City work together to cut down on resource use.

GIS enables solutions

Sustainability is about wise resource use and recycling. It's about clean energy and readily available transportation. It's about cities built for people, not cars. All of which is made much easier with the GIS technologies that have enabled the Masdar team to bring all of these elements to work cohesively together.

Design teams working on Masdar City have stated that the project work and project goals would be infinitely more difficult without GIS. A walk around project offices confirms this, as anyone doing so is likely to see small groups standing around the ArcGIS Explorer, ArcGIS for Desktop, web map, and ArcGIS for AutoCAD monitors to point out issues and discuss alternative solutions. Maps help people make major decisions every day that benefit this project. Even with all that has been done, the project teams know they've only just scratched the surface regarding how they will leverage the city's comprehensive and well-designed GIS to help reach the goals and benefit the residents of Masdar City.

Mackay Regional Council shares data resources and increases productivity with web-based GIS

SECTOR: Regional government

Sandra Janson, Manager of Information Services, Mackay Regional Council, Mackay, Australia

Mackay, Australia, is well known for its sugar and mining industries, as well as its coastal proximity to the Great Barrier Reef, the earth's largest coral reef system. It's also a fast-growing region whose development community and public have a high demand for location-based information. Taking this demand head on, the Mackay Regional Council uses GIS to openly share its spatial data through the web, which has resulted in increased staff productivity and government transparency.

Due to rapid growth in population and development, staff members at the Mackay Regional Council found themselves confronted with an increasing number of requests for GIS data that was stretching their capacity. The council was already using GIS to support several other business operations, and knew that an online GIS solution could help by automating the delivery of information to those demanding it.

MiMAPS brought together all of the data resources within a single portal that delivers critical information related to the region's land use and assets. All of the information is organized and freely available through the web-based GIS application (shown here). Courtesy of Mackay Regional Council.

The council worked with Esri Australia to create and implement what is known today as MiMAPS (Mackay Internet Mapping and Property System). MiMAPS brought together all of the GIS data within a single portal that delivers information about land and property, planning and land use, services and infrastructure, and community facilities. All of the information is neatly organized within an online GIS environment and freely available to those who need it.

Esri's GIS technologies provide the backbone for the integrative solution that pulls together data from multiple business operations into one easy-to-use and multipurpose website. The front end of the application provides a straightforward way for the public and business community to access the information they need, while the back end enables the council to manage and update its data and map services.

By organizing its spatial data resources and serving them through a single GIS portal, the council estimates a 50 percent reduction in the number of calls coming into planners.

By organizing its spatial data resources and serving them through a single GIS portal, the council estimates a 50 percent reduction in the number of calls coming into planners, which is freeing staff for other pressing tasks. Overall, the solution has significantly improved service delivery, guaranteed more accurate information, and made the council more accessible to its clients. MiMAPS is now a well-established solution at the council, and, at the time of this writing, MiMAPS was being upgraded to run on ArcGIS for Server.

Many stakeholders within the regional council are benefiting from the GIS solution, and outside organizations, including developers, planners, real estate agents, solicitors, emergency services agencies, and local shops, are making frequent use of the council's detailed stock of GIS data, proving that in the information age, data is truly a resource.

Benefits Returned by Using GIS to Organize and Manage Data Resources
Saved time—50 percent reduction in the amount of direct requests to planning staff.
Increased productivity—Internal and external users find the information faster, with less hassle.
Improved transparency—There's no down time, information is accessible 24 hours a day, 7 days a week.
Better planning—Developer community's direct access to information permits more accountable planning.

As shown in this summary table, the Mackay Regional Council received several benefits as a result of using GIS to organize and manage its spatial data resources. Courtesy of Mackay Regional Council.

Glynn County, Georgia, uses mobile GIS to manage resources and lower costs

SECTOR: Asset management
P. Hunter Key, GIS Manager, Glynn County, Georgia

Asset mapping is critical to any local government that wants to effectively manage its resources. Just ask Glynn County, in southeast coastal Georgia, where mobile GIS is used to inventory everything from park benches to road signs. It's an approach with tangible returns. For example, by creating a GIS inventory of its streetlights, the county was able to identify hundreds of lights that the local power company was incorrectly billing it for.

The county pays roughly $20,000 a month to power its streetlights. For some time the county suspected that many of the lights it was paying for should actually be billed to local neighborhood associations and school districts. However, there was no easy way to determine this because the billing records used by the power company were over ten years old and stored on hard-copy maps.

Glynn County turned to its GIS department to correct the outdated maps and analyze which lights should and should not be billed to the county. Staff used Esri software to convert the hard-copy maps to a GIS layer stored in a geodatabase. Staff then embedded the data within a simple mobile application that field crews used to verify the data, correct its errors, and map the missing lights. Once the new inventory was completed, it was a simple matter to overlay the streetlight locations with boundary data to assess which lights were rightfully theirs.

The use of Esri's mobile GIS technologies led to an effective and time-efficient solution. The GIS team easily pulled off the entire project on time and on schedule and mapped approximately 2,000 streetlights in less than two weeks.

What stood out most to Glynn County's GIS Manager P. Hunter Key was the agile and cost-effective nature of mobile GIS. With this in mind, his department has repurposed the streetlight solution into a set of mobile applications that field crews now use to map county-owned resources, which, once in the GIS, can be used for an assortment of management needs. On this point, Key said, "The real win for us was not just the savings, but the workflow we could duplicate, which translates to a return on investment in other areas as well."

> The real win for us was not just the savings, but the workflow we could duplicate, which translates to a return on investment in other areas as well. P. Hunter Key, GIS Manager, Glynn County, Georgia.

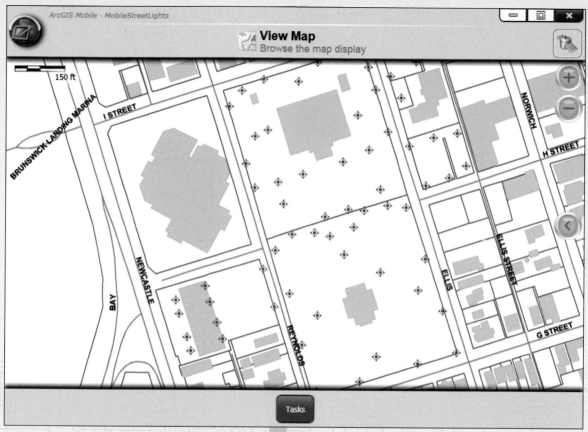

Glynn County, Georgia, uses Esri's ArcGIS for Mobile to inventory and map several different types of county-owned assets, from park hardware to streetlights. Storing this information in GIS helps the county efficiently manage these resources. For example, by creating a GIS inventory of its streetlights (shown here), the county was able to identify hundreds of lights that were being incorrectly billed by the local power company. Courtesy of Glynn County, Georgia.

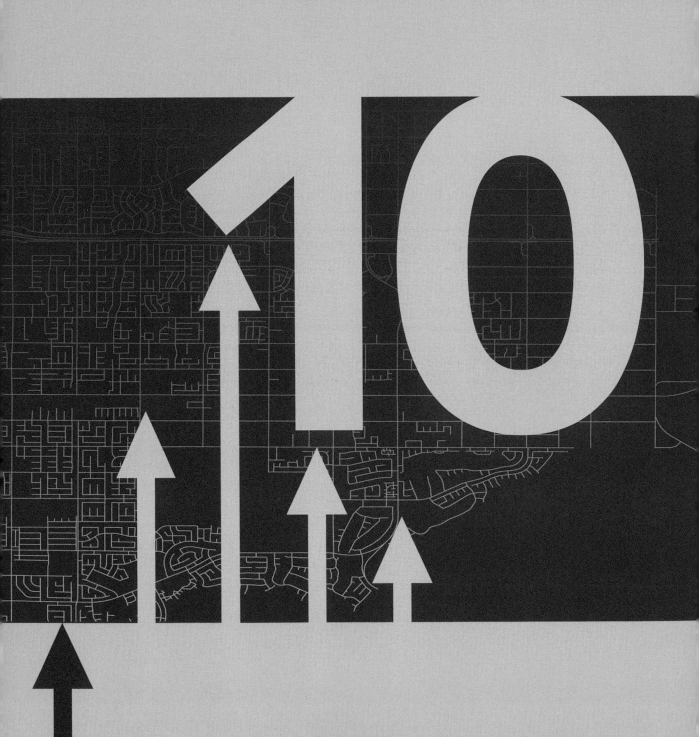

Chapter 10

Aid budgeting

The allocation of budgets to sustain a government's operations is a difficult process even in the best economies. Restrictions are placed on the collection of monies with equally strict guidelines on how those funds are allocated. Customary budgeting methodologies look to allocate funds based on programs and revenue sources. As widespread financial belt-tightening causes many fiscal planners to sharpen their pencils, GIS is helping bring their spreadsheets into focus. With its analytical, reporting, and tracking functions, GIS is a natural "geospreadsheet."

In general, the budget process attempts to match revenues with anticipated costs. The rise in stimulus-related websites provides evidence to the rise of place-based accounting. This methodology provides a platform for government to allocate monies based on geographic and demographic need and are spatially tracked and managed. Additionally, program managers planning their budgets can apply GIS to processes such as fiscal impact analysis, economic feasibility studies, grant justification and verification, reimbursements, and estimating allocations related to ongoing operations.

Traditional accounting practices tie revenues and expenditures to specific activities or customer numbers. Using GIS during budgeting suggests that expenditures of staffing, equipment, property management, and most business operations are spatially related or connected to specific locations. Likewise, any organization or department can use GIS to understand where its revenue is coming from and better allocate budget based on a better understanding of a community's needs. By linking revenues and expenditures to geography, GIS is an exceptional tool for preparing a detailed cost analysis.

While most financial and accounting professionals expect to see budgets in the form of rows and columns, more advanced staff members will recognize that they can perform their budget analyses more efficiently by looking at the geographies or territories assigned to the revenues and expenditures.

Adams County, Illinois, uses GIS to rapidly assess flood damages

SECTOR: Emergency response
Joye Dell Baker, GISP, CFM, Adams County, Illinois

Adams County is located in west central Illinois along the Mississippi River. Periodic flooding is a reality for this county's government and citizens. However, by making a modest investment in GIS, Adams County has dramatically improved its ability to respond to flood events, secure disaster assistance, and ultimately get citizens back into their homes after the water recedes.

> By making a modest investment in GIS, Adams County has dramatically improved its ability to respond to flood events, secure disaster assistance, and ultimately get citizens back into their homes after the water recedes.

The county was an early adopter of GIS. Its initial system was built in 1992 through a cooperative and multiparticipant effort that included the City of Quincy and the five local utility companies operating within the county. From this strong foundation, the GIS was developed to meet the needs of several entities and the public. It was a successful implementation that received national recognition in the GIS industry. It was even used by the Federal Geographic Data Commission as one of the models for building the National Spatial Data Infrastructure (NSDI).

Lessons learned lead to GIS solutions

Adams County's GIS was still in its infancy when massive flooding struck the area during the spring of 1993. The entire floodplain within the county was under water. Homes and businesses were destroyed, and the season's crops were inundated by several feet of water. Public entities claimed more than $9 million in infrastructure damages, and county residents and businesses claimed another $4.7 million.

Much of the area within the floodplain was a total loss, and as the water receded, the county was required by the Federal Emergency Management Agency (FEMA) and the Illinois Department of Natural Resources (IDNR) to complete damage assessments on the remaining structures in the floodplain.

At the time, the county did not have a comprehensive set of tools for completing the damage assessments. The GIS was still in its very early stages, plus permit records were in disarray and the property ownership information was kept on paper in the assessor's office with no simple way to tie the structures' information to their ground locations within the county.

The unorganized records complicated the work of carrying out damage assessments. Needless to say, gathering and organizing the records, then doing the inspections was a massive undertaking. From start to finish, the work took

The flooding that Adams County experienced in 1993 prompted the development of a GIS-based approach for assisting flood response. This system was put into play with great success when massive floods returned. Courtesy of Adams County, Illinois.

twenty-two county employees four months to complete with a total labor cost of $155,443.

The time it took to complete the inspections had a direct impact on county residents, who could not rebuild their homes until their property had been assessed and they had received a new building permit from the county. Most residents and business owners in Adams County had to wait four months or longer for their assessments to be completed and new building permits to be issued. Adding to the wait and the cost was an additional step they had to take to obtain an elevation certificate from an engineering firm that formally defined the base height of the structures on their property.

As bad as the flooding was, and with all the challenges it posed to the county's response teams, it was somewhat fortuitous that the floods occurred at the same time that the county's GIS was burgeoning. The county recognized that many of the challenges it faced when carrying out damage assessments could readily be solved with GIS.

> The county recognized that many of the challenges it faced carrying out damage assessments could readily be solved with GIS.

With flooding still fresh in the minds of local and federal agencies, and a young GIS brewing for more uses, Adams County approached FEMA with the idea of developing a GIS-based approach for future flood responses. FEMA officials liked the idea and granted the county $35,000 to build the GIS database it envisioned.

A GIS solution for assessing flood damages

The county used the FEMA grant to purchase the equipment to develop a GIS-based solution for assessing flood damages. The county then dispatched field crews equipped with GPS and digital cameras to every structure in the floodplain to capture images and their exact locations. While at each site, they also recorded the structure's base elevation so that residents would no longer have to pay for elevation surveys before being issued new building permits.

The data coming in from the field was organized using Esri software to create a GIS layer of building structures within the floodplain. This GIS-based structure inventory was then tied to other information such as floodplain maps, elevation certificates, previous damage assessments, building dimensions, floor plans, and building permits.

The project took two full-time staff members in the field and one in the office approximately three months to complete. Once finished, the county had a comprehensive inventory of all the structures in the floodplain, and each

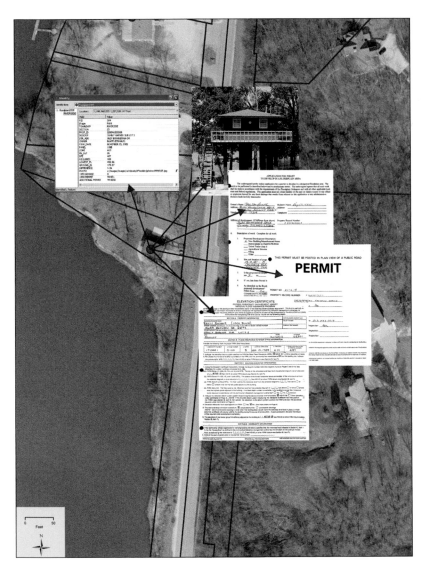

Adams County, Illinois, uses GIS to assist in flood response activities. Key to the solution is an inventory of all the structures in the floodplain. As shown in this image, users can employ the GIS to locate the structures on a map, then access other important information such as photos, building permits, and previous damage assessments. Courtesy of Adams County, Illinois.

structure in the inventory was linked to a set of critical attributes that described it. The county also joined the GIS inventory to the permitting process, which meant that any new buildings or improvements would automatically get added to the GIS inventory as new permits were issued, thus keeping the inventory up to date and ready to use for the next wave of flooding.

GIS-ready when flooding returns

Widespread flooding returned to Adams County in 2008. Unlike the floods from more than a decade earlier, this time the county was able to quickly turn out maps and data critical to local responders as well as state and federal emergency management agencies.

The GIS made it easy to identify what structures were flooded along with the total value of these properties. It also made it possible for the county to stay aware of the full scope of the flooding as it unfolded. The mapping and reporting capabilities of the GIS were used to monitor the flooded area, assist evacuation processes, and generate the budget figures used to request federal disaster assistance.

Once again, the county was called on to perform detailed damage assessments after the water receded. This time, county officials had a complete inventory of the structures in the floodplain, and no guesswork was required to determine what had been flooded and where to send crews. The county completed the damage assessments in less than three weeks with only seven employees at a total cost of just $14,313, which was roughly $140,000 less than the 1993 amount. The savings were

> # The GIS made it easy to identify what structures were flooded along with the total value of these properties. It also made it possible for the county to stay aware of the full scope of the flooding as it unfolded.

great, but possibly more important was the fact Adams County was able to speed the budgetary processes required to secure disaster assistance and get people back into their homes and businesses.

GIS is a point of pride for Adams County, Illinois. The county was an early adopter of the technology, and today uses GIS for a variety of applications, from highway maintenance to crime analysis. Numerous case studies could be written that describe the return on investment the county gets back from its GIS, but probably no number can be estimated for the value it provides when used, as Adams County did, to help get citizens out of harm's way and back into their homes after disaster strikes.

Assessing Flood Damages, with and without GIS			
	Staff	Time	Costs
Prior to GIS (1993 Flood)	22 employees	4 months	$155,443
With GIS (2008 Flood)	7 employees	3 weeks	$14,313
GIS Benefits:	15 fewer employees required	Overall process shortened by 3.75 months	$141,130 saved

As shown in this table, the integration of GIS into damage assessment workflows led to substantial savings in staff, time, and money, while at the same time helping to rapidly generate budget figures used to secure federal disaster assistance. Courtesy of Adams County, Illinois.

The City of Moreno Valley, California, analyzes foreclosures using GIS

SECTOR: Real estate
Stephen Jarrett, GIS Specialist, City of Moreno Valley, California

Moreno Valley, California, lies sixty-five miles east of Los Angeles and is home to approximately 193,000 residents. The city and its surrounding region experienced a rapid pace of residential development during the early 2000s, but when the national housing market faltered later in the decade, Moreno Valley found itself—like much of the nation—facing a growing number of property foreclosures. As part of its response, city officials used GIS to identify areas of Moreno Valley most vulnerable to foreclosure, then used these results to help secure $11 million in federal foreclosure assistance.

City officals used GIS to identify areas of Moreno Valley most vulnerable to foreclosure, then used these results to help secure $11 million in federal foreclosure assistance.

During the height of a national foreclosure crisis, the US Department of Housing and Urban Development (HUD) allocated nearly $4 billion in emergency relief to help local governments maintain the health of their neighborhoods and local economies. As part of HUD's application process, municipalities seeking assistance were required to show which areas within their jurisdictions they would target with the aid. It was essentially a map-based question, and one well suited to a GIS-based response.

To help fulfill HUD's application requirements, City of Moreno Valley GIS Specialist Stephen Jarrett gathered input data from internal and external sources, then used ArcGIS for Desktop to map the location of all the foreclosed, vacant, or abandoned properties. This work generated a GIS layer storing 1,653 property locations under some form of foreclosure. Jarrett used overlay analysis to summarize the foreclosed properties by census tract. This summary data was then examined in relation to areas within the city where neighborhood stabilization programs already existed. For the final step of the process, staff used GIS to delineate the neighborhoods most threatened by foreclosure and therefore most in need of assistance.

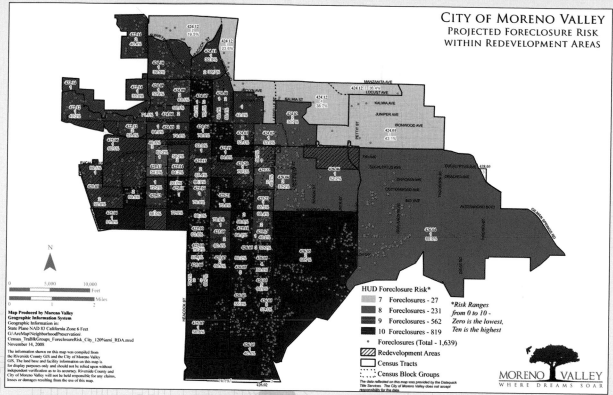

CITY OF MORENO VALLEY
PROJECTED FORECLOSURE RISK WITHIN REDEVELOPMENT AREAS

HUD Foreclosure Risk*

7	Foreclosures - 27
8	Foreclosures - 231
9	Foreclosures - 562
10	Foreclosures - 819

Risk Ranges from 0 to 10 - Zero is the lowest. Ten is the highest

- Foreclosures (Total - 1,639)
- Redevelopment Areas
- Census Tracts
- Census Block Groups

The data reflected on this map was provided by the Dataquick Title Services. The City of Moreno Valley does not accept responsibility for the data.

MORENO VALLEY
WHERE DREAMS SOAR

N

0	5,000	10,000
		Feet

0	1	2
		Miles

Map Produced by Moreno Valley
Geographic Information System
Geographic Information in:
State Plane NAD 83 California Zone 6 Feet
G:\ArcMap\NeighborhoodPreservation\
Census_TraBlkGroups_ForeclosureRisk_City_120%ami_RDA.mxd
November 14, 2008

The information shown on this map was compiled from
the Riverside County GIS and the City of Moreno Valley
GIS. The land base and facility information on this map is
for display purposes only and should not be relied upon without
independent verification as to its accuracy. Riverside County and
City of Moreno Valley will not be held responsible for any claims,
losses or damages resulting from the use of this map.

The City of Moreno Valley, California, used GIS to map foreclosure incidents within the city and project future risks from the problem. The maps from the analysis, such as the one shown here, were used to help secure several million dollars in federal relief, which helped mitigate the impacts caused by the crisis. Courtesy of Stephen Jarrett, GIS Specialist, City of Moreno Valley, California.

The GIS analysis made it possible for city planners to analyze foreclosure patterns and predict where the problem would concentrate over an eighteen-month period. The maps and information generated by Jarrett's GIS analysis went directly into Moreno Valley's application to HUD for emergency relief funds, and ultimately helped the city secure millions of dollars in much needed federal assistance, which the city then put to immediate use in its efforts to mitigate the damaging impact of foreclosures on local residents and neighborhoods.

The City of Redlands, California, uses GIS to budget services and cover new costs

SECTOR: City management
Philip Mielke, GIS Supervisor, City of Redlands, California

Determining how to cover the long-term costs associated with new shopping centers and housing projects is an important consideration of any city concerned about the long-term sustainability of its communities. One case in point is the city of Redlands, California, which has 68,000 residents and is located sixty-three miles east of Los Angeles. The city used GIS to help design an equitable fee system for new development that is now offsetting the cost for a much needed police station by $6.5 million.

> The City of Redlands, California, used GIS to help design an equitable fee system for new development that is now offsetting the cost for a much needed police station by $6.5 million.

Cities often receive an influx of development proposals when the economy is strong. These attractive proposals and the revenues they promise must be carefully considered by city government, because when the economy slows and revenues dwindle, the municipality is left to pay the bill for standard services like street maintenance and law enforcement. It's an issue Redlands takes head on by charging fees for new development—fees that then get allocated to support the services and infrastructure these new developments require over time.

To predict future service demands and thus determine development fees, Redlands used GIS to create map layers from historical incident data for police, fire, public works, and utilities operations. This work produces what GIS Supervisor Philip Mielke refers to as the "money out maps." Desktop GIS software is then used to overlay the incident maps with the city's land-use layer to create a spatial data product from which GIS staff summarized and reported the total volume of services provided by land-use type.

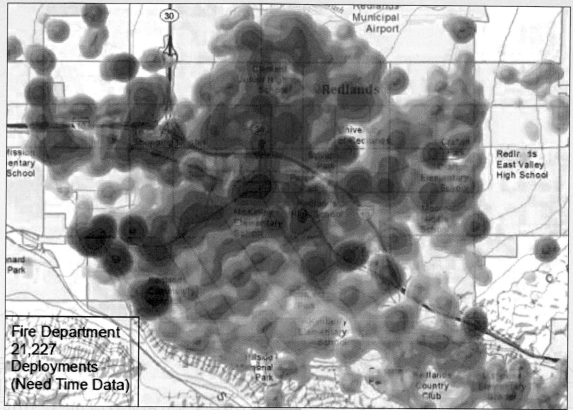

Fire Department
21,227
Deployments
(Need Time Data)

The City of Redlands uses GIS to map the patterns and volumes of city services, such as law enforcement and fire protection (shown here). The spatial data behind these maps has been leveraged to calculate a development fee schedule based on historical service demands by land-use type. These fees are then levied on new development projects to offset the costs for the facilities and services they will require over time. Courtesy of City of Redlands, California.

The GIS analysis used by Redlands to determine service demands has provided a type of data-driven intelligence with many practical uses within the city. Department managers now use the data and maps to visualize how staff resources are being used and to assist in planning decisions and resource allocation. The summary statistics were also used to develop a fee schedule for new developments based on the type of new land use proposed. By taking this approach, and charging fees based on a realistic assessment of future demands, GIS has helped Redlands budget for and offset the costs of capital improvement projects required to support the long-term service demands associated with new development; it's a case in which GIS is helping Redlands stay true to its motto as "a city that works."

Summary

GIS returns on its investment

For the organizations and governments that use it, GIS has evolved into a mainstream technology because it continues to yield a return on investment. Quantifiable benefits include time saved to perform a task, costs avoided or money saved, revenue generated or recovered, and increased productivity and accuracy. Other key payoffs include decision support, increased efficiency, and improved communication.

GIS works better when organizations implement an enterprise-wide system that connects geography to the business process. With online services, such as interactive mapping, agencies can better serve citizens, businesses are enhanced, and internal government operations—IT infrastructure, data management and warehousing, information exchange, and field force automation—are more efficient and flexible.

Measuring the benefits of GIS

Data creation and maintenance set the foundation for moving forward with a successful GIS. A strategy of thorough documentation and criteria analysis, coupled with recent breakthroughs in mobile GIS software that enable field-to-office and Government 2.0 applications, are paving the way for organizations as they aggressively strive to provide more efficient and effective services and products.

As these processes move forward and more GIS applications are put into use, GIS users will look for ways to identify and document how the technology benefits their organizations and their customers and how they can add to their return on investment by integrating GIS into other business workflows. Reflecting on how these business tasks would be accomplished without a GIS can bring to mind many of its benefits.

Continuously measuring the benefits of GIS can be contagious, and organizations and businesses will begin to find new ways of calculating and tracking benefits and exceeding those benefits each year. Users, managers, and public officials will recognize the benefits and business value of GIS to justify continued investments in GIS.

The future promises that the benefits of GIS will be measured with more sophisticated and powerful formulas that will enable us to identify more advantages for all. Measuring the benefits of GIS will positively transform organizations into more effective and successful operations.

To learn more about the benefits of GIS or to submit your own benefits of GIS case study, visit www.esri.com/measuringbenefits.

Case study credits

Chapter 1: Save time

GIS keeps information flowing to response agencies and public during Queensland flooding
Adapted from Keith Mann (Esri writer), "Cloud GIS: Fast, reliable support for disaster response," *ArcUser,* Spring 2011, http://www.esri.com/news/arcuser/0311/cloud-gis.html; courtesy of Esri Australia.

GIS delivers results to Alameda County Registrar of Voters
Adapted from "GIS Delivers Results to Alameda County Registrar of Voters," *Government Matters,* Spring 2009, 6–7; courtesy of Alameda County, California.

Louisiana Army National Guard deploys GIS to make the most of its data
Adapted from "The Mission of Coordinating Safety," *Public Safety Log,* Fall 2009, 1–3; courtesy of Louisiana Army National Guard.

Chapter 2: Save money

GIS erases the cost of graffiti in Riverside, California
Adapted from Matt Keeling, "Tracking the True Cost: ArcGIS Server-based tool mitigates graffiti," *ArcUser,* Summer 2009, 16–17; courtesy of City of Riverside, California.

GIS-assisted permitting saves Honolulu money and generates new revenue
Adapted from "Big Impression: Honolulu leverages GIS for better services citywide," *More than Maps,* A Government Technology Solution Spotlight: Esri, 2009, 3–4; courtesy of City and County of Honolulu, Hawaii.

Saving livestock saves millions
Adapted from "New Zealand's Animal Health Board Fights Bovine TB," *GIS for Agribusiness,* Summer 2009, 4; courtesy of the New Zealand Animal Health Board.

Chapter 3: Avoid costs

Baltimore County, Maryland, strategic plans reveal millions in return on investment from GIS
Adapted from Matt Freeman, "Baltimore County's GIS Strategic Business Plan Reveals a Significant Return on Investment," *Government Matters,* Winter 2009, 8–9; courtesy of Baltimore County, Maryland.

Bonner County, Idaho, manages invasive weeds with GIS
Adapted from Cori Keeton Pope, "Bonner County, Idaho, Manages Invasive Weeds with GIS," *Government Matters,* Fall 2009, 6, 10; courtesy of Bonner County, Idaho.

Los Angeles Bureau of Sanitation uses GIS to avoid costs and improve workflows
Adapted from "GIS Keeps Local Governments Going: Los Angeles Dodges Unneeded Expenses with a Smart Routing Solution," *ArcNews,* Spring 2010, 22; courtesy of City of Los Angeles, Bureau of Sanitation.

Chapter 4: Increase accuracy

Charting the roads that connect the vast Navajo Nation
Adapted from "Charting the Roads of the Vast Navajo Nation," *ArcNews,* Fall 2011, 30–31; courtesy of Navajo Division of Transportation.

City of Alpharetta, Georgia, uses GIS to get an accurate census count
Courtesy of City of Alpharetta, Georgia.

Hudson, Ohio, increases government transparency and workflow accuracy using GIS
Courtesy of City of Hudson, Ohio.

Chapter 5: Increase productivity

GIS-based asset management peaks productivity for Colorado Springs
Adapted from "GIS Creates a Paperless Asset Management System in Colorado Springs," *Government Matters,* Spring 2009, 8, 10; courtesy of City of Colorado Springs, Colorado.

Collaborative Utility Exchange maps Johnson County utilities
Courtesy of Johnson County, Kansas.

GIS-based work order management increases productivity for the Consolidated Utility District
Courtesy of Consolidated Utility District, Rutherford County, Tennessee.

Chapter 6: Generate revenue

Pueblo County, Colorado, grows economy with GIS
Adapted from "Pueblo County, Colorado, Grows Economy with GIS," *Government Matters,* Winter 2009, 1, 10; courtesy of Pueblo County Colorado.

Westfield, Indiana, gets fair revenues from GIS
Courtesy of City of Westfield, Indiana.

Woodstock, Georgia, adds revenue by using GIS to analyze storm water billing
Courtesy of City of Woodstock, Georgia.

Chapter 7: Increase efficiency

Golden, Colorado, does more with less by using GIS for sign management
Courtesy of City of Golden, Colorado.

Irish councils automate data collection with GIS to improve public feedback processes
Adapted from "Dún Laoghaire-Rathdown and Fingal County Councils: Sharing ideas and pooling resources to improve services for citizens," An Esri Ireland Case Study, 2010; courtesy of Fingal County Council.

Mobile GIS improves code enforcement services in McAllen, Texas
Adapted from "Mobile GIS Improves Code Enforcement Services in McAllen, Texas," *ArcNews,* Fall 2011, 22–23; courtesy of City of McAllen, Texas.

Chapter 8: Automate workflows

The Virginia Department of Forestry uses GIS to automate workflows in office and the field
Adapted from "Virginia Department of Forestry: Extending Enterprise GIS to Virginia Department of Forestry's Mobile Staff," An Esri Case Study Brochure, 2008; courtesy of Virginia Department of Forestry.

EastLink Tollway: GIS puts road construction project in the fast lane
Adapted from "EastLink Tollway Project: GIS Puts Road Construction Project in the FastLane," An Esri Case Study Brochure, 2009; courtesy of Jason Clark.

Nova Scotia Power saves $200,000 in labor costs through GIS field connectivity
Adapted from "Nova Scotia Power Saves $200,000 in Labour Costs through GIS Field Connectivity," *ArcNorth News,* Spring 2011; courtesy of Esri Canada.

Chapter 9: Manage resources

Masdar City relies on GIS to help create one of the world's most sustainable urban developments
Adapted from "Masdar City: The World's first carbon-neutral city," A Business Case Study from Esri UK, 2009; courtesy of Derek Gliddon, GIS Manager, Masdar City, Abu Dhabi.

Mackay Regional Council shares data resources and increases productivity with web-based GIS
Adapted from "Mackay Regional Council: MiMAPS,"An Esri Australia Case Study Brochure; courtesy of Mackay Regional Council.

Glynn County, Georgia, uses mobile GIS to manage resources and lower costs
Adapted from "Glynn County Utilizes Mobile Data Collection for Street Light Inventory," A Glynn County Case Study Brochure; courtesy of Glynn County, Georgia.

Chapter 10: Aid budgeting

Adams County, Illinois, uses GIS to rapidly assess flood damages
Courtesy of Adams County, Illinois.

The City of Moreno Valley, California, analyzes foreclosures using GIS
Courtesy of the City of Moreno Valley, California.

The City of Redlands, California, uses GIS to budget services and cover new costs
Courtesy of City of Redlands, California.